Human Knowledge
and
Human Nature

HUMAN KNOWLEDGE
AND
HUMAN NATURE

A new introduction to an ancient debate

PETER CARRUTHERS

OXFORD UNIVERSITY PRESS
1992

Oxford University Press, Walton Street, Oxford OX2 6DP

Oxford New York Toronto
Delhi Bombay Calcutta Madras Karachi
Petaling Jaya Singapore Hong Kong Tokyo
Nairobi Dar es Salaam Cape Town
Melbourne Auckland

and associated companies in
Berlin Ibadan

Oxford is a trade mark of Oxford University Press

Published in the United States
by Oxford University Press, New York

British Library Cataloguing in Publication Data
Data available

Library of Congress Cataloging in Publication Data
Carruthers, Peter, 1952–
Human knowledge and human nature : a new introduction to an
ancient debate / Peter Carruthers.
p. cm.
Includes bibliographical references and index.
1. Knowledge, Theory of. 2. Innate ideas (Philosophy)
3. Empiricism. I. Title.
BD161.C352 1992   121—dc20   91–23735
ISBN 0–19–875101–X
ISBN 0–19–875102–8 (pbk.)

Typeset by Pentacor PLC
Printed in Great Britain by Biddles Ltd,
Guildford & King's Lynn

Contents

for Isaac
no blank slate he

Preface

THIS book began life as a series of lectures that were never delivered. In the spring of 1988 I was invited to take part in the second Sino-British Summer School in Philosophy, due to have been held in Beijing in August 1989. The idea was to produce written texts of the lectures, both to be made available to those attending the Summer School, and for later publication in Chinese translation. I began work that July, and had the lectures completed by the following May. But then the session of the Summer School was unfortunately cancelled, following the suppression of the pro-democracy demonstrations in Tiananmen Square by the People's Liberation Army on 4 June 1989.

My original brief was to write on British Empiricism for an audience of Chinese academics and graduate students whose English might not be perfect, and who, while having some knowledge of empiricism and of Western philosophy generally, would have had little experience of active work within our tradition. Accordingly, I was to teach not only by content but by example. That spirit has survived into the present book. Not only have I tried to provide a clear and succinct introduction to many of the issues in the theory of knowledge particularly relating to empiricism, but I have also attempted to make an original contribution to the subject, thereby illustrating how serious work in philosophy can at the same time be accessible. I also had the idea that my lectures might form a useful model of how one can take past traditions of thought seriously while at the same time retaining a critical distance—trying, from our contemporary perspective, both to rework the insights and to avoid the errors of history. That spirit, too, has survived into the present book. I am grateful to Nick Bunnin, as Chairman of the British Committee of the Summer School, for extending me the original invitation. The texts of my lectures were prepared under the title 'Empiricism, Nativism and A Priori Knowledge'.

I am grateful to the following individuals for their advice, and for criticisms of earlier drafts: David Archard, Angela Blackburn, Nick Bunnin, Bob Hale, Síle Harrington, Susan Levi, Cynthia Macdonald, Alan Millar, David Smith, Nicholas White, Tim Williamson, Xiao Yang, and two anonymous readers for Oxford University Press. I am also grateful to my students at the Universities of Essex and of Michigan, who were obliged to study earlier drafts of the book as a text, and who gave me the benefit of their objections and—equally important—their frequent incomprehension.

Some of the ideas in this book have been delivered as talks at the University of East Anglia, the University of Michigan, the 1990 Joint Session of the Aristotelian Society and Mind Association, and Trinity College Dublin. I am grateful to all those who participated in the subsequent discussions for their comments and criticisms.

Some of the material in Chapters 9 and 10 has been previously published under the title 'What is Empiricism?' in the *Proceedings of the Aristotelian Society* (supplementary volume 64, 1990). I am grateful to the Editor of the *Proceedings* for permission to make use of it.

One final note: throughout this book I shall use the colloquial plural pronouns 'they' and 'their' in impersonal contexts, in place of the masculine singular 'he' and 'his' required by strict English grammar. I believe that grammar needs changing in this respect, since it contrives to give the appearance that only men ever do or think anything worth mentioning.

P.C.

Sheffield
April 1991

1

Introduction: Modes of Knowledge

THIS book is about the theory of knowledge, focusing especially on debates between empiricism and various forms of rationalism. In this first chapter I shall outline the nature of my project, which is to reassess the claims of classical empiricism from our present perspective.

Problems of Knowledge

We can distinguish three dimensions along which theories of knowledge may differ from one another. There can be dispute about the *extent* of human knowledge, about the *sources* of human knowledge, and also about what knowledge itself *is*. I shall say something briefly about each of these three dimensions in turn.

First, philosophers may disagree with one another about the extent of human knowledge, about *how much* we may be said to know. At one end of the scale are various kinds of sceptic, who claim that we know very little, or at least a good deal less than we think we do. Extreme sceptics claim, for example, that we can know nothing beyond our own current states of consciousness—that is, our own present thoughts and experiences. At the other end of the scale are various sorts of realist, who maintain that we know a good deal about ourselves and the world around us. Between these two poles there is space for many intermediate positions. One that may be worth mentioning in particular is phenomenalism. Phenomenalists allow that most of our ordinary beliefs about physical objects do constitute knowledge, but only under a particular (non-realist) interpretation. They claim that such beliefs do not really concern a world of objects existing independently of our minds, but relate only to recurring patterns within our experience. When I speak of the chair on which I am sitting, for example, they claim that I am really only referring to a familiar bundle of experiences, which is apt to recur periodically within the stream of my consciousness. So while I do perhaps know that the chair exists, and may continue to know this even in its absence, this is not really knowledge of anything outside my own mind.

Secondly, there may be disagreement about the possible sources of human knowledge, about the avenues through which we may hope to obtain knowledge. At one extreme is the position adopted by classical empiricists such as Locke and Hume, who claimed that the only source of substantive knowledge is experience (understood broadly to include both memory and introspection). At the other extreme is the view of Plato, who held, on the contrary, that experience cannot yield genuine knowledge, since it concerns states and objects that are constantly changing. He claimed that the only source of true knowledge is the human intellect, which may obtain for us knowledge of the unchanging world of Forms, or universals. In between these two, lies the position of classical rationalists such as Descartes and Leibniz, who maintained that knowledge may be obtained *both* through experience *and* through the use of pure intellect. Plato's view has been endorsed by almost no one since, and need not detain us here. It derives from his peculiar conception of the nature of knowledge. But we shall be very much concerned with the dispute between classical empiricism and classical rationalism. Indeed, it will form the main focus of this book.

Finally, philosophers may disagree about what knowledge itself *is*, about how the concept of knowledge should properly be defined. This issue has not loomed very large in the work of most theorists of knowledge since Plato, at least until very recently. But in our own era a variety of accounts have been proposed. All are agreed that in order to count as knowledge something must at least be believed and be true—you cannot know that grass is green unless you believe that grass is green, and unless it is in fact true that grass is green. All are equally agreed that there is more to knowledge than mere true belief. Differences arise as to what the third component of knowledge might be. Some hold that it is justification, some that it is causation by the fact that is believed to obtain, some that it is causation by a reliable process. We shall return to these issues in some detail in Chapter 5.

It is usual for books on the theory of knowledge to begin with a discussion of scepticism. I propose, however, to reserve my consideration of this issue until last. For how we handle it may depend very much upon what we think about the nature of knowledge, and its possible sources. Moreover, there are, in any case, a good many issues that can be discussed without referring to the problem of scepticism. These are probably best treated first. For in my experience, when someone has once been introduced to the problem of scepticism it is hard to persuade them to take seriously any other issue in the theory of knowledge. It is true, however, that any conclusions we may reach can only be provisional. For if, in the end, the extreme sceptic cannot be answered, then it may turn out that we can know nothing about any of these other issues.

What is distinctive about contemporary debates in the theory of knowledge is that they mostly bypass altogether the question of the possible sources of human knowledge. All the attention is devoted to what knowledge itself is, and to attempts to defend or to undermine various forms of scepticism. It will be one of the themes of this book that such a concentration of effort is a mistake. In my view, the correct definition of knowledge is a good deal less significant than is sometimes thought. And it should be obvious that the question of the extent of human knowledge simply cannot be answered without considering its possible sources. On this issue (as, arguably, on many others) we can only make progress by returning to the historical roots of our current debates. Accordingly, I shall now say something further about the dispute between empiricists and rationalists concerning the question of how knowledge may be obtained, focusing initially on how empiricism itself should properly be characterized.

Empiricism: An Initial Sketch

Empiricists have defended two distinctive negative theses about the sources of human knowledge, which may or may not be intimately connected with one another (I shall leave this open for the moment). First, they have been opposed to any form of nativism, for example denying that any concept or any knowledge is innate (inborn). Secondly, they have denied that we may obtain substantive knowledge of the world a priori, insisting rather that all such knowledge must be grounded in experience. (The notion of the a priori can be understood in either of two distinct ways—most usually, as knowledge which can be arrived at by a process of thought alone, or, alternatively, as knowledge which is not learned from experience. The difference is not at present important, but will later prove so.) I shall shortly say something about each of these strands, returning to them in greater detail in subsequent chapters. But first I want to make some remarks about further candidates for inclusion in our initial characterization of empiricism.

Most empiricists have also been foundationalists, maintaining that the architecture of our knowledge consists of a super-structure supported by foundations. They have held that some of our beliefs, particularly reports of immediate experience or sense-data, have a privileged position with respect to the others, providing them with their ultimate support. (In contrast, coherentists hold that knowledge may be characterized as a set of mutually coherent and supportive true beliefs. While foundationalists picture knowledge in the shape of a building, standing on a foundation, coherentists picture it as a sort of net—perhaps spinning unsupported in deep space—held together by the internal tensions between its elements.) But foundationalism can hardly have been distinctive of empiricism, since all classical rationalists, too, have shared a similar commitment to it. Descartes, for example, held that all knowledge must be founded on the data of immediate experience together with simple truths of reason. Foundationalism may turn out to be a necessary ingredient in empiricism, however, even if it is by no means distinctive of it. This possibility will be left open for now.

A more plausible candidate for inclusion in our initial account of empiricism is a commitment to phenomenalism, since most empiricists have believed that all our knowledge must in the end reduce to knowledge of patterns in our own subjective experience. But there are two things wrong with this suggestion. The first is that it ignores the position of Locke, who certainly thought that we could have knowledge of objects outside ourselves.[1] Admit-

[1]   See his *Essay Concerning Human Understanding*, bk. IV, ch. XI.

tedly, his arguments for this view are weak, and most later empiricists have believed that he was inconsistent in failing to embrace phenomenalism, given his general claims about the genesis of concepts and of knowledge. Be that as it may, we can hardly maintain that a commitment to phenomenalism is fundamental to his position, given his own explicit rejection of the doctrine.

The second thing wrong with the above suggestion is that phenomenalism is best viewed as a consequence of the three strands already mentioned, rather than as a distinct concern in its own right. For the rejection of nativism means that all our beliefs must in the end be decomposable into concepts that can be derived from experience. This, together with foundationalism, then accords the data of immediate experience a central position within the overall structure of our knowledge. And when this is combined with opposition to substantive a priori knowledge, we appear to have the implication that our knowledge can reach no further than patterns within the data of immediate experience itself. I shall return to consider the possible connection between phenomenalism and anti-nativism in more detail in chapter 11. For the moment the issue may be left to one side.

One further candidate for inclusion in our initial characterization of empiricism would be a commitment to the imagist theory of thinking. On this view, all thoughts consist of sequences of mental images, and all concepts come to represent their objects in virtue of resembling them. While Locke's commitment to imagism is half-hearted, and does no serious work within the argumentative structure of the *Essay*, in the writings of Berkeley and Hume it appears to be more significant. In *The Principles of Human Knowledge*, for example, Berkeley argues from imagism as a premiss to the conclusion that it is impossible for us even to think about an independent material reality, claiming that an idea can be like—can resemble—nothing but another idea, and that it is impossible for us to image an unexperienced object. Hume, too, seems to endorse such a position.[2]

It is doubtful, however, whether imagism is really independent of the two negative strands in empiricism already mentioned, particularly opposition to nativism. For the major attraction of imagism, for empiricists, is that it provides them with an apparently plausible theory of concept acquisition. If concepts

---

[2] See his *Treatise of Human Nature*, bk. i, pt. ii, sect. vi and bk. i, pt. iv, sect. ii.

are images—mental pictures—then it is easy to see how they might be acquired through experience, by a process of *copying*. On this account, the concepts we deploy in thinking will be simply reproduced aspects of experience, laid down earlier in memory. We shall return to empiricist theories of concept acquisition in some detail in chapter 4. For the moment, the point is that there is no reason to include a commitment to imagism within our primary characterisation of empiricism. Rather, that commitment is best understood as a natural consequence of the empiricist thesis that all concepts must be acquired from experience.

Representation and Truth

We have just now touched upon issues that have been the subject of much recent debate amongst philosophers of mind and of language, concerning the nature of representation and of truth. What all are now united on is that the imagist theory of thought is hopelessly inadequate, but there is agreement upon little else. Some strands in this debate need to be mentioned briefly here, if only to be set to one side.[3]

One issue of note concerns the contrast that can be drawn between coherence conceptions of truth on the one hand, and correspondence conceptions on the other. While coherentists about knowledge hold (as we saw briefly above) that the justification for a belief consists in its membership of a suitably coherent set of beliefs, coherentists about truth go further. They maintain that for a belief to be true just *is* for it to be a member of a suitably coherent set. Correspondence theorists, on the other hand, maintain that for a belief to be true is for it to correspond to the facts. Minimally, they maintain that for a belief or statement to be true, there must be some fact—some state of affairs—in virtue of which it is true. The idea here is that our beliefs have to measure up to an independently given reality. I intend to presuppose a correspondence conception of truth

---

[3] For an introductory exposition of the way in which issues in the theory of meaning may have bearing on the theory of knowledge, see J. Dancy, *An Introduction to Contemporary Epistemology* (Oxford: Blackwell, 1985).

throughout the course of this book, for reasons that I shall now briefly explain.

An obvious difficulty for coherence conceptions of truth is that it then seems quite possible that a belief might be true at one time but false at another, without there being any relevant change in its subject-matter. For example, the belief that the Earth is flat formed part of a coherent and mutually supportive set of beliefs in the fifteenth century, but no longer does so today. Does this then mean that it *was* once true that the Earth is flat, but that it is so no longer? Coherentists appear to have two possible avenues of response. One is to claim that representation, and hence belief-content, are also a matter of coherence. On this account there will be no genuine conflict between our beliefs and those of the fifteenth century because, through being embedded within different coherent sets, the beliefs they assert are not the same as the ones we deny. The other option is to introduce a considerable degree of idealization into the account of coherence, for example claiming that to count as true a belief must belong to a set that incorporates and explains all possible past and future observational data.

Neither of the above options appears at all attractive. It seems undeniable that I do contradict the beliefs of the fifteenth century by asserting that the Earth is round. And even if we can imagine what it would be like to be in possession of a maximally coherent theory incorporating all possible observational data, it seems to remain an open question whether or not that theory would be true—whether it would match up to the facts. Indeed, the coherence conception of truth (like phenomenalism, as we shall see in chapter 11) is best understood as a retreat from an initial realism designed to outflank scepticism, by *defining* truth as what our best (most coherent) theory would describe. But if it is possible to argue against scepticism directly, as I shall suggest in chapters 11 and 12 that it is, then the coherence theory of truth will be inadequately motivated.

Even amongst those who endorse some version of a correspondence theory of truth, there remain disputes about the extent to which truth is genuinely objective. Realists about truth maintain that it is, each one of our beliefs about the world being determinately true or false irrespective of whether we are capable (either in practice or in principle) of discovering its status. Anti-realists, in contrast, maintain that truth is constrained by relations of epistemic accessibility—in its strongest version, claiming that there is no truth except verifiable truth;[4] in weaker versions, claiming that there can only be truth where there can be evidence that bears on truth.[5] Since this is a dispute arising out of considerations in the theory of meaning (or of representation generally), it would take us too far out of our way to discuss it properly here. But my own position is that it is in virtue of our conceptual capacities—our abilities to identify and classify items in the world—that our thoughts come to represent what they do;[6] and that our best theory of our conceptual capacities, when embedded within our best theory of the material world and our relation to it, is one that supports a realist conception of truth. I shall return to the matter briefly in chapter 12. For the most part I shall simply assume, for purposes of argument, that truth is indeed objective. Our question is whether we have *knowledge* of any truths, and if so, how—that is, whether we may be realists about knowledge as well as realists about truth.

Two Strands in Empiricism

Let me now return to consider the two main strands within the empiricist tradition in a little more detail. Empiricists, as I say, have been united in their opposition to nativism, whether concerning innate knowledge, innate beliefs, innate concepts, or innate mental structures embodying information about particular domains of knowledge, such as language or vision. They have insisted, on the contrary, that all our knowledge and beliefs must in one way or another be derived from experience; that all concepts must either be abstractable from experience or definable in terms of concepts that can be so abstracted; and that all the psychological mechanisms involved in learning are general ones, the same for all domains of knowledge. Here there

---

[4] See A. J. Ayer, *Language, Truth and Logic* (London: Gollancz, 1936) for the classic defence of such a position.

[5] Michael Dummett has defended a position of this sort. See many of the papers in his *Truth and Other Enigmas* (London: Duckworth, 1978). See also Crispin Wright, *Realism, Meaning and Truth* (Oxford: Blackwell, 1986).

[6] On this see my *Tractarian Semantics* (Oxford: Blackwell, 1989), ch. 10.

is sharp conflict with those who belong to the rationalist tradition, including Plato, Descartes and Leibniz (as well as contemporary writers such as Noam Chomsky and Jerry Fodor). Much of our attention in this book will be focused on this debate, especially in chapters 4–8.

While empiricists have been opposed to nativism in any of the above forms, it is worth stressing that they have not rejected all forms of innateness. On the contrary, they have characteristically held that the division of the mind into distinct faculties (for example, into sensation, imagination, and understanding) is innately given. They have also held that the basic psychological processes involved in the various modes of acquiring our beliefs are innate, such as principles of association amongst our ideas. (Indeed, they are surely obliged to hold such a view. For one cannot, from nothing, learn how to learn.) Some empiricists have even held views that can be described as *naturalistic*, relying upon aspects of an innately given human nature to provide part of their suggested solution to some problem. For example, Hume's position concerning our belief in the continued existence of physical objects is naturalistic in this sense, since he holds that, given our nature, we cannot but believe that items such as tables and chairs continue to exist unperceived.[7] But this is not to ascribe to us innate knowledge of the physical world. Nor is it to claim that the faculty through which we acquire this belief (the faculty of imagination, in Hume's view) is structured in such a way as to contain innate information about the world. His idea is simply that the general laws of human psychology ensure that we shall have beliefs concerning the continued existence of physical objects, irrespective of sceptical arguments to the contrary.[8]

Empiricists have also been united in denying that we can attain substantive knowledge of the world a priori. They have insisted, on the contrary, that all substantive knowledge is empirical, needing to be grounded in experience. They are then immediately presented with a challenge. Namely, to provide some account of our knowledge of logic and mathematics, neither of which appears to be at all empirical. I shall now spend

[7] See the *Treatise*, bk. I, pt. IV.
[8] On this see P. F. Strawson, *Skepticism and Naturalism* (London: Methuen, 1985).

some time outlining and discussing the manner in which empiricists have attempted to reply to this challenge.

The majority have chosen to respond via a particular interpretation of the subject-matter of those disciplines, such as mathematics, which are a priori. They have claimed that the truths of logic and mathematics are *analytic*, being concerned only with internal relations between our concepts (ideas) themselves. So the knowledge that we have of these truths does not concern any aspect of the world independent of our minds, except in so far as our concepts may have been derived from experience of such a world. Here empiricism may properly be contrasted with platonism. For the platonist believes, on the contrary, that the propositions of logic and mathematics concern an abstract (non-physical and changeless) but genuinely mind-independent realm of objects, including universals such as beauty and wisdom, as well as mathematical entities such as the natural numbers (1, 2, 3, etc.) and the perfect square. This debate, too, will occupy us considerably in the chapters that follow, especially in chapters 2 and 3, and again in chapter 10.

Some empiricists have been inclined to disparage analytic truths as being trivial and uninformative. This is a definite mistake. Plainly some analytic truths are trivial—neither 'Everything is identical with itself' nor 'It is either raining or not raining' are likely to be news to anyone. But many are not, only being recognized for the first time after the collaborative labour of many generations of thinkers. Indeed, a great many important scientific discoveries have only been made possible by conceptual or mathematical advance. This is why I elected to characterize the present strand in empiricism by saying that it consists in denying that there is *substantive* a priori knowledge of the world, rather than by saying that it is the denial of *informative* a priori knowledge. (By knowledge that is substantive, I shall understand knowledge which is *either* contingent—as is our knowledge of our own states of mind and our own existence—*or* which concerns entities that exist independently of the human mind— as does mathematical knowledge, given a platonist account of the subject-matter of mathematics.) In my view the debate between empiricists and platonists ought properly to concern the subject-matter of logic and mathematics, not its usefulness or cognitive significance.

Some later empiricists (notably Mill and Quine)[9] have adopted a rather different manner of responding to the challenge provided by our knowledge of logic and mathematics. They have denied that such knowledge (indeed any knowledge) is really a priori. On this view, the propositions of mathematics, like all other propositions, are contingent and empirical—they could possibly be false, and are derived ultimately from experience. But they exist at a high level of abstraction and generality, very far removed from the impact of experience. It is this that gives rise to the illusion that they may genuinely be known independently of the course of our experience, and that they obtain necessarily, in all conceivable situations. Rather, such truths, like any others, are grounded in experience and revisable in principle in the light of future experience. It is just that their connection with experience is not easily noticed, due to their abstractness.

Quine's famous image of our beliefs as constituting a web, only connecting with experience directly at the periphery, provides a graphic illustration of this idea.[10] We are to picture our beliefs as forming an interconnected network, some of whose members (the periphery) are direct reports of experience, whereas others (such as propositions of logic) are very far removed from it. Changes at the periphery occasion alterations elsewhere in the network, none of our beliefs being immune from possible revision. Which of our beliefs should in fact be revised in face of the changing course of our experience will be a matter of what adjustments would provide us with the most satisfying overall explanation. The truths of logic, although in principle revisable, are hardly ever altered because of the massive changes this would occasion elsewhere in the network. There will almost always be more economical ways of accommodating recalcitrant experiences.

The trouble with such an account, however, is to explain how any changes in belief may rationally be required of us, if all beliefs are equally empirical. If all our beliefs face the tribunal of experience together, then there is nothing to give substance to the idea that some adjustments of belief are rationally demanded of us in the face of other alterations elsewhere in the network.

---

[9] See J. S. Mill, *A System of Logic* (London: Longmans, 1879) and W. V. Quine, *From a Logical Point of View* (Cambridge, Mass.: Harvard University Press, 1953).
[10] See 'Two Dogmas of Empiricism', in *From a Logical Point of View*.

On the contrary, all changes in the network of belief will be merely causal, it being a purely psychological matter what changes will take place within a subject's beliefs in response to any given experience. In order for the idea of a rationally required change to make sense, some at least of our beliefs must be held constant, functioning as norms that mediate changes amongst the others. This is precisely the role traditionally accorded to the a priori propositions of logic and mathematics.

However, to insist that propositions of logic and mathematics function as norms, and are hence not empirical (not learned from experience), is not necessarily to grant that their status is inviolable. What an empiricist should say about this matter will depend upon whether or not they think that analytic truths are objective. Some, like Ayer, have held that they are, reflecting mind-independent relations between concepts (whose existence, in turn, is not platonic but mind-*dependent*).[11] Others, following the later Wittgenstein, would deny this, holding that a proposition achieves a priori status somewhat as follows. Finding ourselves psychologically incapable of seeing how a given proposition (such as '$15 + 17 = 32$') could be false, we adopt the convention of elevating the results of such limitations to the status of a norm of description, using it henceforward to mediate changes between other beliefs. But this status need not be inviolable: what is a necessary truth (a norm) at one time, may cease to be so at another. I shall not pursue these ideas any further in this book.[12] Suffice it to say that even granting that there exists a class of analytic truths, there remain difficult questions concerning the objective or subjective status of such truths.

We have canvassed two ways in which empiricists, who deny that any substantive knowledge may be obtained a priori, might respond to the challenge provided by logic and mathematics. One is to maintain that our knowledge within these domains, while a priori, is analytic. The other is to deny that our knowledge of logic and mathematics is genuinely a priori. Their only remaining option is apparently to take the radical step of

[11] See Ayer, *Language, Truth and Logic*, ch. 4.
[12] Those interested may like to read Crispin Wright, *Wittgenstein on the Foundations of Mathematics* (London: Duckworth, 1980), pt. III. See also my *Metaphysics of the Tractatus* (Cambridge: Cambridge University Press, 1990), chs. 4, 11 and 15.

denying that any propositions of logic or mathematics are *true*. Few empiricists have taken this course, for obvious reasons. But it has been defended recently by Hartry Field.[13] I shall return to consider this option briefly in chapter 2. But I shall mostly assume that the correct strategy for an empiricist is to allow that logic and mathematics are a priori, and contain many known truths, but to deny that they are *substantive*—that they give us knowledge of anything that exists independently of the human mind.

Having said something about the questions that belong to the theory of knowledge in general, and the nature of the empiricist movement in particular, it is now time to outline what I plan to do in this book.

The Project

The basic aim of this book is to reconstitute and defend the essentials of the empiricist tradition from within our contemporary perspective. The major question to be asked is whether anyone today can be justified in being an empiricist, or whether, on the contrary, empiricism has now been decisively superseded. I shall be arguing for the former of these options. This is despite the fact that one of the two main strands within classical empiricism (namely, the denial of nativism) has come under increasing pressure in recent years (and rightly so, as we shall see in later chapters). I shall be arguing that the core of empiricism, if properly characterized, remains as vibrant and defensible today as it ever was. My aims are thus both to provide a particular interpretation of the nature of the empiricist tradition and also to defend that tradition (the latter partly via the former).

While I shall occasionally comment upon specific classic texts, I shall mostly take the broad outlines of their authors' views for granted. It should be emphasized, however, that I shall not take empiricism to be defined at the outset by any particular set of ideas and doctrines—and certainly not by the two main strands already distinguished. For a good deal of what is at issue here is

[13] See his *Science without Numbers* (Oxford: Blackwell, 1980) and *Realism, Mathematics and Modality* (Oxford: Blackwell, 1989).

what constitutes the true core of empiricism. This is a matter, not of explicit doctrines, but of fundamental motivations.

It may be objected that the term 'empiricism', as generally understood, may simply be *defined* as the doctrine that all knowledge must be grounded in experience. So questions of motivation arise not at all, and it follows immediately that empiricism is inconsistent with nativism. With this I disagree. While our grasp of empiricism may be partially characterized by the claimed connection between knowledge and experience, it seems to me undeniable that it is also partly governed by our acquaintance with the views defended by those generally known as 'empiricists'. (An analogy from literature: while our understanding of the term 'romanticism' may bring to mind a particular definition, it is also partially characterized by our knowledge of the works of those generally described as 'romantics'.) It is this text-based strand in the notion of empiricism that becomes primary when we raise the question of our own relationship to traditional empiricism. In this sort of case what we wish to know is whether our own views are or are not in conformity with the basic concerns of earlier empiricists. Our question is the extent to which we may count ourselves as belonging to the same tradition of thought.

Rather than understand empiricism to be defined by any particular set of doctrines, I shall take the central texts of Locke and Hume as representative of the motivation behind empiricism in general, particularly the former's *Essay* and the latter's *Treatise* and *Enquiry Concerning Human Understanding*. For I think that if these authors had never written, one would never have thought of seeing enough in common between philosophers such as Bacon, Hobbes, and Berkeley to describe them all as 'empiricists'; but that the reverse is not the case.

The issue before us may usefully be considered in terms of the notions of *core* and *periphery*. What we shall need is to distinguish between those aspects of the empiricist tradition that constitute its core (that are genuinely essential to it), and those that are merely peripheral (that could be given up without losing anything of fundamental importance). This is a matter of discovering what it was that the classical empiricists were most concerned to establish or show. For if, as I shall argue, the empiricist opposition to nativism turns out to be a definite mistake, then we shall need to know whether this aspect of the tradition may simply be cut away from the remainder, leaving a coherent body of doctrine intact that is still recognizably empiricist in spirit. We shall therefore need to know which, if either, of the two main strands in classical empiricism is the more fundamental, or whether there might be some further characterization of the core of empiricism that would explain its commitment to both.

Successful completion of my project will thus require two things. First, an assessment of the truth or falsity of the two main doctrines that have traditionally been defended by empiricists. This task will occupy us through the major part of this book. Second, we shall need an account of the essential core of empiricism that will embrace what is true, while allowing us to exclude what is false. This will form the topic of chapters 9 and 10, although I shall also say something briefly about the matter shortly.

The course of the argument will proceed roughly as follows. In chapters 2 and 3 I shall consider the empiricists' rejection of substantive a priori knowledge, focusing particularly on their rejection of platonism. In chapter 2 I shall consider the case that can be made out in support of platonism. In chapter 3 I shall consider both how an empiricist might try to undermine that case and their positive reasons for thinking platonism to be false. This will be the most technical chapter of the book, and readers are advised to take it slowly. My conclusion will be that the grounds for the empiricist rejection of substantive a priori knowledge are indeed powerful.

Then in chapters 4–8 I shall consider the empiricist rejection of nativism. In chapter 4 I shall discuss Locke's arguments against innate knowledge and innate concepts. In chapter 5 I shall consider whether the very idea of innate knowledge is incoherent, by virtue of our conception of what knowledge itself *is*. In chapter 6 I shall consider the case provided by contemporary cognitive science for innate information-bearing mental structures, concentrating on Chomsky's arguments for an innately structured language faculty. In chapter 7 I shall consider the case for saying that there are innate concepts, considering particularly the arguments of Fodor. Then in chapter 8 I shall consider the case for saying that we possess substantive innate knowledge, in particular of truths concerning our own psychology. My conclusion will be that the empiricist case against nativism is lamentably weak, and that at least some of our knowledge and concepts are very probably innate.

Given this situation—with one strand in classical empiricism endorsed and one rejected—we then face the question of the fate of the movement as a whole. In particular, we need to know whether the denial of nativism is truly an essential part of the empiricist project. In chapters 9 and 10 I shall argue that it is not. Rather, the core of empiricism consists in a certain sort of naturalism (distinct from that briefly characterized above). The main concern is to insist that claims to knowledge should only be endorsed where we can begin to give an account, in terms of natural processes, of the manner in which that knowledge might have arisen in us. The reason why the early empiricists rejected nativism, I shall argue, is that the only form of account available to them at the time, of how a belief or concept might come to be innate, was a non-natural one, namely direct intervention by God. But now, with the advent of evolutionary theory, we can remain true to the empiricist project while at the same time embracing nativism.

In the final two chapters I shall explore the consequences of this result for the problem of scepticism, treating separately the problem of induction and the problem of the external world. Here I shall argue that, given commitments to certain sorts of nativism, solutions to these problems may readily be constructed. Indeed, it will turn out that not only the classical empiricists' rejection of nativism, but also their tendency towards scepticism or phenomenalism, were historical accidents. Each resulted from the scientific ignorance of the time, rather than from any essential aspect of the empiricist enterprise. The final position will be an empiricism denuded of its opposition to nativism (while still retaining the classic opposition to substantive a priori knowledge), endorsing a robust realism about our knowledge of the physical world. This position, I argue, is not only distinctively empiricist, but correct.

2

Knowledge out of Reason: Platonism

In this second chapter I shall consider arguments supporting one of the major forms of rationalism, namely platonism with respect to the truths of mathematics, logic, and conceptual analysis.

Rationalism and Platonism

Rationalists are united in believing that it is possible for us to acquire substantive knowledge of the world a priori. Platonism is (almost always)[1] a form of rationalism, since platonists believe that we may obtain a priori knowledge of a realm of abstract objects, including numbers and universals. This knowledge is genuinely substantive, since such objects are held to exist independently of the human mind. But platonism is not the only form of rationalism. Many rationalists have also believed that we can know a priori of the existence and immortality of the soul, of the existence of an all-powerful, good God, and of the freedom of the will. Indeed, in Locke's own day many believed that we could know a priori the basic constitution of the natural world around us. This was one form of the medieval–Aristotelian conception of science. It was believed that the various natural sciences form deductive systems (somewhat like geometry), the axioms of which are self-evidently true as well as innately known. (This was the form of rationalism that Locke himself was most concerned to attack directly.)

While platonism is only one form of rationalist belief, it nevertheless holds a central position in the debate with

---

[1] Quine and his followers believe that sets and other abstract objects may be known empirically, on the basis of total science.

empiricism. This is because logic and mathematics represent the only examples of a priori knowable truths that are relatively uncontroversial. Almost everyone (whether empiricist or rationalist) agrees that it is true that $2 + 2 = 4$, and that we may know this a priori, by a process of reasoning alone. So if it could be shown that this truth is genuinely substantive, it would follow that rationalism is vindicated: it would follow that it is possible for us to acquire substantive knowledge through reason alone. It would then be hard to see why there should be any objection in principle to the use of reason discovering for us other substantive truths, such as the existence of the soul or the freedom of the will. In contrast, in all other domains where rationalists have sought substantive a priori knowledge—for example, attempting to construct proofs of the existence of the soul, or of the existence of God—empiricists have believed (in most cases rightly) that they could detect errors in the reasoning used that were quite independent of the empiricist–rationalist debate. Such arguments thus do not constitute a problem for empiricism, in the way that the truths of logic and mathematics manifestly do. Nor do they amount to very powerful evidence of the truth of rationalism.

I therefore propose to concentrate upon platonism in this chapter. This is not only because of its crucial position in the overall debate, but also because this is the point at which the issues between empiricists and rationalists can be presented in their sharpest focus. In chapter 10 I shall return again to consider rationalists' claims to other forms of substantive a priori knowledge.

Platonists maintain that alongside the physical world there also exists a realm of abstract (non-physical and changeless) entities, peopled by universals (such as greenness, friendship, and beauty), and by mathematical objects (including numbers and geometrical figures). They believe that it is these entities that the propositions of logic, mathematics, and conceptual analysis are *about*. They hold that these entities constitute the subject-matter of our logical and mathematical propositions. Platonists also generally believe that abstract entities have a mode of existence which is *necessary*, believing that numbers, for example, exist at all times and in all conceivable circumstances (in all possible worlds). They believe that no matter how distant the past or future one considers, and no matter how different the natural world might have been from the way it actually is, it would still have been the case that numbers exist, and each number would still have had the very same mathematical properties.

Some platonists have held that abstract entities have an existence that is *timeless* rather than necessary. They have held that, so far from existing at all times in all possible worlds, it *makes no sense* to say of an abstract object that it exists at a particular time, or at all times—just as it makes no sense to say of an abstract object that it exists in a particular place, or in all places. But in my view this thesis lacks intuitive plausibility. For sentences like 'The number 7 always has and always will exist' just do not have the obvious kind of senselessness possessed by such sentences as 'The number 7 is everywhere'. (Everyone will concede that abstract objects are *spaceless*.) On the contrary, such sentences seem perfectly intelligible, if not obviously true. However, nothing of any great significance turns on this issue for our present purposes. In what follows I shall confine attention to that version of platonism which claims necessary existence for abstract entities. For it is easier to express platonist doctrines, and to present arguments in their support, if platonic objects are supposed to have necessary rather than timeless existence.

When it comes to explaining our knowledge of abstract entities, many platonists have believed that we possess a special faculty of intellectual intuition, modelled by analogy with sense-perception. The purpose of this faculty is to make available to us facts about the abstract realm, just as our senses serve to make available to us facts about our immediate physical environment. The idea is that when we come to believe simple conceptual or mathematical truths, we *intuit* an aspect of the abstract realm— we 'see it with our mind's eye'. So the fact that the abstract realm contains the number 7, which has the property of being prime, is to explain my belief that 7 is prime in something like the manner in which the fact that there is a cup on the desk in front of me explains my perceptual belief that there is a cup on the desk. In both cases the state of affairs in question is supposed to exert an influence upon my mind, via faculties of intuition and perception respectively.

Does this then mean that the belief that 7 is prime is not really a priori? Should we think of intellectual intuition as a way of acquiring experiential (empirical) knowledge of the abstract realm? I think not. There are two reasons for this. The first is that, in contrast with genuine sense-experience, there are no sensations characteristic of acquiring a new mathematical belief. When I see that there is a cup on the desk, my experience has its own distinctive phenomenological character. But when I 'see' that 7 is prime, all that happens is that I come to have a new belief. Phenomenologically speaking, it remains the case that the only process involved in the acquisition of new mathematical beliefs is thought. The second difference between intellectual intuition and genuine perception is that the supposed faculty of intellectual intuition remains a bare hypothesis, lacking independent confirmation. In connection with genuine sense-experience, in contrast, we have well-grounded beliefs about the structures, locations, and modes of operation of the various senses.[2]

What are the main arguments that can be given in support of platonism? I shall begin by considering the arguments for believing that there are abstract entities at all, taking the case of numbers first (it is here that the arguments are strongest). I shall then discuss the existence of universals. Finally, I shall consider the arguments for believing, not only that there are abstract entities, but that such entities have mind-independent, indeed necessary, existence.

Numbers as Objects

A powerful case can be made for saying that numbers exist as abstract individual objects. It has been presented most clearly by the nineteenth-century platonist and logician, Gottlob Frege, though many of the points were implicit in the writings of Plato.[3] The argument begins by noting that in much of our

---

[2] For further discussion of the proper characterization of the senses, and of the way in which we may know how many senses there are, see the papers entitled 'Categorising the Senses' by Mark Leon and Norton Nelkin, in *Mind and Language*, 1988 and 1990 respectively.

[3] See Frege's little masterpiece *The Foundations of Arithmetic 1884*; (trans. J. Austin, Oxford: Blackwell, 1950), from which much of what follows in this section is derived. See also Plato's *Republic*, bk. vi.

thought and speech we treat numbers *as* individual objects. Indeed, in most of their occurrences in sentences numerals (number-words) appear to function just like proper names. It is worth spelling this out in some detail, distinguishing six different respects in which language seems to treat numbers as individual objects.

First, we predicate things of numbers, just as we predicate things of physical objects. Compare the sentence 'David is a small man' with the sentences '7 is a small number', or '7 is prime'. Just as in the first sentence the use of the name 'David' serves to refer to a particular human, of whom we then predicate the property of being small, so too in the other sentences it appears that the numeral '7' is referring to an individual object, of which we then predicate the property of being small, or the property of being prime. In short: the sentence '7 is prime' is of subject–predicate form, with the word '7' serving to introduce its subject, just as a proper name does.

Secondly, numbers can be referred to by means of definite descriptions (phrases of the form 'the such-and-such'), which apparently serve to pick out one individual thing from others. Thus compare the two sentences 'The tallest living person is the oldest' and 'The successor of 6 is prime'. In the first sentence the definite description 'The tallest living person' serves to pick out and distinguish one from the rest of us (even if we do not know whom), who is then claimed to be the oldest. Similarly in the second sentence, the phrase 'The successor of 6' seems to pick out and refer to a particular number (in fact the number 7), which is then said to be prime.

Thirdly, we quantify over numbers (making 'some' and 'all' statements about them). This suggests that they form a domain of individual things, just as people form a domain of individual things. Thus we may say 'Some number between 6 and 10 is prime', just as we might say 'Some person in the room has fair hair'. The most natural way to understand the truth-conditions of such sentences is that they are true if and only if there is some individual—nameable—thing in the domain having the property in question. Then it seems that numbers, like people, must be individual nameable things.

Fourthly, we take statements about individual numbers to imply existence-statements, just as we do statements about people. Thus, in the same way that 'David is in the room' implies 'Someone is in the room' (by virtue of our intention that the name 'David' should refer to an existing individual), so too does '7 is prime' imply 'Some number is prime'. This suggests that we must be intending the numeral '7' to refer to an existing individual thing, in such a way that statements involving that numeral can only be true if the intended referent really does exist.

Fifthly, we make statements of identity concerning numbers, just as we do concerning individual physical objects. Thus just as we might say 'Thatcher is none other than Margaret' (or 'Thatcher and Margaret are one and the same person'), so too we may say '2 + 2 and 4 are one and the same number', or, more simply, '2 + 2 = 4'. (In order to provide a context for the first of these statements, suppose you know that Mrs Thatcher was the British Prime Minister in 1989 without knowing what she looks like, and are then introduced to someone at a reception as 'Margaret' without realizing that she was the 1989 Prime Minister, only later discovering that they are one and the same person.) In this respect, too, language treats numbers as individual objects, which can be either identical or distinct from one another.

Finally, it is a general truth about individual objects that they do not have contradictories, whereas properties do. Nor do numbers have contradictories (which are also numbers).[4] All properties have contradictories, in the sense that for any property, there is another property that is true of something if and only if the first is not true of it. For any property F, there is the property of being not F. The corresponding principle does not hold for objects. It is not the case that for any object there is another object that possesses a property if and only if that property is not possessed by the first object. Rather, almost any two objects will have *something* in common. Similarly, any two numbers will have something in common. For example, both 7 and 640 are larger than 6. We certainly cannot guarantee that for any number, there is another number that lacks just the properties the former possesses, and vice versa.

---

[4] If numbers are treated as properties of sets, so that, for example, the number 7 is identified with the property of having 7 members, then there will also be the property of *not* having 7 members. But this will not itself be a number.

While there are many respects in which our language treats numbers as objects, with numerals functioning in sentences just like proper names, this is not universally the case. In statements of number such as 'Jupiter has 4 moons' the numeral '4' seems rather to occur as an adjective. It appears to be qualifying the noun 'moons', in something like the way that the adjective 'large' does in the sentence 'Jupiter has large moons'. Yet there is a difference. For from 'Jupiter has large moons' we may infer 'Each of the moons of Jupiter is large'; whereas from 'Jupiter has 4 moons' we certainly cannot infer 'Each of the moons of Jupiter is 4'. So if '4' occurs here as an adjective, it is by no means an ordinary adjective.

In fact the quasi-adjectival use of numerals can easily be rendered consistent with their otherwise namelike use, if we notice that the statements of number in which they figure can be rewritten as identities without any loss of content. The sentence 'Jupiter has 4 moons' says the same as 'The number of moons of Jupiter is none other than 4', in which '4' does explicitly occur in the role of a proper name. (Note, in contrast, that we cannot easily say 'The size of the moons of Jupiter is none other than large'. This seems nonsensical.)

Since we use numerals in the manner of proper names, our language treats numbers themselves as individual objects. Then if any statements of arithmetic are true, it seems that numbers must really exist. For compare: the statement 'David is a small man' cannot be true unless the name 'David' does succeed in referring to someone, that is, unless the person David really does exist. Similarly, then, '7 is prime' cannot be true unless the number 7 really exists. For this is a statement about the intended referent of the name '7', just as 'David is a small man' is a statement about the intended referent of the name 'David'. We then apparently face a stark choice: either to accept that numbers are genuinely existing objects, or to reject all statements of arithmetic as false. If the latter is too high a price to pay, then we must accept that numbers exist as individual things.

From the fact that numbers are individual objects, it does not immediately follow that they are *abstract* (non-physical and changeless) objects. But this further thesis may be established with very little additional argument. For it is obvious that numbers are not physical things, like tables and chairs. Thus it really does seem to be nonsensical to ask *where* the number 7 might be, as we noted above. Even those who maintain that numbers are really sets of sets of things (for example, identifying the number 3 with the set of all trios of things) are forced to concede this. For sets of physical things (let alone sets of such sets) are not themselves physical things. Even if each fair-haired person is physical, the set of such people is not.[5]

Moreover, since the truths of arithmetic are necessary, holding with respect to all times and all possible worlds, it is clear that numbers must be changeless things. We certainly have no use for the idea that there are circumstances in which 7 would not have been prime, nor for the idea that at some future time it might cease to be so. What has to be acknowledged, however, is that numbers can undergo change in their relational properties; that is to say, in the relations in which they stand to changing physical things. For suppose Jupiter loses one of its moons. Then at one time the number of its moons is 4 and at a later time it is 3. But this is best viewed as a change in the relations obtaining between numbers and the things in the world satisfying a certain description, rather than a change in the numbers themselves.

Are any Mathematical Statements True?

At this point we should consider how plausible it would be to try resisting the above argument for platonism about numbers by rejecting all mathematical statements as false. This is the option defended by Hartry Field. He agrees with the platonist interpretation of the subject-matter of arithmetic, accepting that '2 + 2 = 4' purports to state a fact about abstract objects. But because he does not accept the existence of such objects, he does not accept that arithmetic contains any truths—just as someone who denies the existence of David would deny that 'David is a small man' can express a truth. Yet he hopes to explain how arithmetic can be useful in its applications, in terms of its *consistency*. In Field's view mathematical statements might have

[5] For a contrary view, see Penelope Maddy 'Physicalistic Platonism', in A. Irvine (ed.), *Physicalism in Mathematics* (Dordrecht: Kluwer, 1990).

been true (had there existed any numbers), although they do not happen to be so; and this is claimed to be sufficient to explain the practical usefulness of such statements.

An initial objection to Field is that his position cannot be adequately motivated. This is because the main difficulty with platonism, as we shall see in chapter 3, is that it cannot provide an acceptable explanation of our knowledge of mathematics. Thus in order to argue that the platonist interpretation of mathematics is false, we shall suppose that mathematics itself consists largely of truths, and indeed is largely known to do so. Such an argument cannot be pursued by those, such as Field, who wish to deny that mathematics consists of truths. Field's attempt to deny platonism by rejecting all mathematical statements as false would then appear to be self-defeating.

In fact Field can reply to this objection. His view is that our theory of the world—our total science—is a better theory if it excludes platonic objects than if it includes them. And one of the main reasons why belief in platonic objects would make our overall theory worse is that we should be left incapable of explaining how we could have knowledge of those objects. So the argument against platonism, that it cannot adequately explain how we could have knowledge of mathematics, can be presented without presupposing that any of our mathematical statements are true. On the contrary, the reason why our theory of the world is better for denying that any mathematical statements are true is that we could not explain how we could have knowledge of them if they were.

However, it is clear that Field's strategy for rejecting platonism is one of very high risk. For it needs to be shown that a science without any mathematical truths *is* a better total theory than a science that includes such truths. To begin with, it is a matter of considerable technical complexity to show that science *can* be presented in such a way as not to presuppose that any mathematical statements are true.[6] But even then, Field's task is not complete. For explanatory completeness is only one good-making feature of a theory. Another is simplicity. So if, as seems likely, scientific theories will be more complex when reformulated so as not to presuppose mathematical truth, it will still not be settled that we should reject mathematics as false. It may be

[6] See e.g. the paper by A. Urquhart in Irvine, *Physicalism in Mathematics*.

that the best overall option will prove to be the platonist one, even accepting that we cannot then explain the genesis of mathematical knowledge.

Perhaps the main objection to Field's position, though, is simply that it is hugely counter-intuitive. It is very hard indeed to induce oneself to believe that '2 + 2 = 4' might in fact be false. As a result, the position contains a crucial dialectical weakness. This is because our pre-theoretical belief in the truths of mathematics is much more firmly grounded than any merely *philosophical* argument for the platonist interpretation of the subject-matter of those truths, such as was sketched in the previous section. Since it requires an argument to convince us that '2 + 2 = 4' is really about abstract objects, it will always be more reasonable for us to think that something has gone amiss with the philosopher's argument than to give up our mathematical beliefs.

Field also faces a related problem. For mathematics is very naturally taken to consist of statements that, if true, are true necessarily—holding good with respect to all possible worlds. But then if those statements are false, they will be necessarily false—which conflicts with Field's claim that they are consistent (true with respect to some possible world). In fact it is almost equally hard to see how '2 + 2 = 4' might be contingent (true of some worlds but not others) as it is to see how it might be false.

Now the objection here is not that if Field were right, it would then be wholly inexplicable why there should happen to be no abstract objects.[7] For Field may reply that the non-existence of numbers is a brute contingency, comparable in status to the claim that there are no physical objects, made by those who have become convinced of the non-existence of matter. (It is obvious that such a person cannot explain *why* there are no physical objects, except perhaps theologically, by saying that God failed to create any. Why then should Field be under any obligation to explain why there happen to be no numbers, other than by saying that God never made any?) Rather, the objection parallels the one made above. It is that our pre-theoretical conviction that mathematical statements are necessary (as

[7] For pursuit of this line of objection to Field, see Bob Hale, *Abstract Objects* (Oxford: Blackwell, 1987), ch. 5, as well as his 'Nominalism' in Irvine, *Physicalism in Mathematics*.

opposed to contingent) is likely to be more firmly grounded than any philosophical argument for a platonist interpretation of the subject-matter of mathematics.

Field's position would seem to be one of last resort for an empiricist. Indeed, it is so barely believable that if it were the only option available, then this might be a powerful reason for embracing rationalism. But in fact there are other possibilities open to us, as we shall see in the next chapter. The conclusion thus far is that the existence of numbers as abstract individual objects can at least be supported by powerful arguments, if not proved outright. At this point it appears that an empiricist who wishes to oppose platonism faces a formidable task.

The Existence of Universals

The case for believing in abstract universals has to be made somewhat differently. For universals are supposed to be what the predicative expressions of natural language (such as 'wise' or 'small') refer to. And we cannot claim that such expressions function in sentences as proper names (as we claimed above that numerals do) without destroying altogether the distinction between name and predicate. It is true that most predicates admit of nominalizations, which can then figure in the subject position in a sentence. Thus, as well as sentences like 'David is wise' we have sentences such as 'Wisdom is hard to acquire'. But it is doubtful whether anything of much significance depends on this. For it is generally easy to see how the content of sentences containing such nominalizations may be expressed without loss by sentences containing only the corresponding predicate. Thus what 'Wisdom is hard to acquire' really says is that it is hard to become wise.

What does seem to require us to recognize the existence of universals is this. We accept existence-claims that are not claims about the existence of individual objects (as is 'There is something that is human'), but are rather claims about the existence of the properties those objects possess. Thus consider the sentence 'There is (there exists) something that we all are (namely human)'. This surely expresses a truth. It seems undeniable that there is something common to us all. In which case it appears that we must not only accept the existence of individual humans, but also of what they have in common—the universal feature humanness. It is also true that there is (there exists) something that grass and leaves have in common, namely greenness; and so on. If we accept such truths, we appear to be accepting the existence of universals.

From the fact that universals exist, it does not immediately follow that they are abstract. Nothing has as yet ruled out the idea that they may rather be immanent in the physical world (as Aristotle maintained). Thus it may be that universals such as greenness are, in part, physically present in the individual things that instantiate them (that is, in the individual leaves and blades of grass that share the property of being green). This sort of view has been defended recently by David Armstrong.[8] He argues that we need to postulate identical natures in things (immanent universals) if we are to give an adequate explanation of the fact that different things can share similar causal powers, and of what laws of nature are. He thinks that the best explanation for the fact that all water behaves similarly in similar circumstances—freezing at zero centigrade, dissolving sugar, and so on—is that all water shares an identical inner constitution that necessitates that it behave as it does. This inner constitution is a repeatable feature of the world, being identically present in many different items of water, and is hence a universal. But it is, on this view, a universal that is *present in* the physical items that instantiate it.

What apparently forces us to accept the abstractness of universals is that we accept existence-claims with respect to them even when they are *not* instantiated in the physical world. For example, it appears to be true that there is (there exists) something that *nothing* is (namely, a unicorn, or a dragon). If this is accepted, then it seems to follow that universals, in existing independently of their instances (where they have instances), are not physically present in the natural world. Their changelessness may then be established using similar considerations to those deployed in the case of numbers. For we have no use for the idea that the universals dragonness or greenness

---

[8] See his books *Universals and Scientific Realism* (Cambridge: Cambridge University Press, 1978) and *What is a Law of Nature?* (Cambridge: Cambridge University Press, 1983).

themselves undergo change. Although a withering blade of grass is at one time green and at a later time not, this would best be viewed as a change in the relation in which it stands to the universal greenness, rather than a change in that universal itself.

We may conclude that the existence of universals as abstract entities can also be supported by powerful considerations, just as can the existence of numbers as abstract individuals. Then a priori conceptual truths such as 'Anything red is coloured' and 'No surface can be red all over and green all over at once', in being concerned with internal relations between universals, will give us a priori knowledge of a realm of abstract entities. But this is not yet to say that such knowledge will be genuinely substantive. For it is important to note that we have not yet established that universals (or indeed numbers) have mind-independent existence, let alone that their existence is necessary (that they exist at all times in all possible worlds).

In order to see that there is a gap here waiting to be bridged, consider what might naturally be said about the existence of *sentences* as abstract types. Given that both the vocabulary and the rules of sentence formation for a language have been fixed, it is natural to think of the sentences of that language as already existing, independently of their being tokened in speech or writing. And this is just what we say: there are many sentences of English that no one has ever expressed, or perhaps ever will. Those sentences may remain unexpressed because they are too complicated ever to be uttered, or too silly; or simply because no one ever happens to think of them. Nor do we have any use for the idea that a sentence-type might undergo change. The sentence 'David is wise' is as it is; it cannot change without becoming *another* sentence. Yet we are unlikely to be tempted to think of sentences as existing independently of the rules implicit in the practices of those who speak the language. On the contrary, if those rules had never arisen in the way that they have (for example, in the case of English, if the Normans had never invaded and conquered England), then the sentences in question would never have existed.

It is thus natural to think of sentences as abstract but mind-dependent entities, depending for their existence upon the rules implicit in the practices of the native speakers of a language. A similar position is then possible in connection with numbers and universals. One might concede that they are abstract entities (being non-physical and changeless throughout the time of their existence), but maintain that their existence supervenes upon the existence, and properties, of the human mind. Note, however, that I do not take myself to have *shown* that sentences have mind-dependent existence. My point has only been to establish that such an idea makes sense, and hence to show that more needs to be done if we are to establish full-blown platonism with respect to numbers and universals. Some have maintained that sentences, too, have necessary existence, and I do not claim to have answered them here.[9]

Objective Truth and Objective Necessity

Are there any reasons for believing that numbers and universals are abstract entities that have an existence independent of the human mind? The main argument for such a claim is premised upon our intuitive belief in the objectivity of truths about such entities. Thus we are inclined to think that the truth of 'There is something that nothing is (namely a dragon)' is wholly independent of our beliefs about the matter. Even if we were all to believe that there are dragons, it would still be true that there is something that nothing is (namely a dragon). Indeed, not only are truths about universals independent of our beliefs, but we are strongly inclined to think that they are independent of our very existence as well. Even before there were any human beings, it was still true that there was something that nothing was (namely a dragon). Equally, even if there had never been any human beings, it would still have been true that there is something that nothing is (namely a dragon). In which case it would seem that universals themselves (the entities such truths are about) must exist independently of the human mind.

Similar arguments can be deployed in support of the mind-independent existence of numbers. Indeed, in this case the impression of objectivity is, if anything, even stronger. The truth of '7 is prime' is, we think, quite independent of anything

    [9] See, for example, Jerrold Katz, *Language and Other Abstract Objects* (Oxford: Blackwell, 1981).

we might believe about the matter. Even if we all believed otherwise—either through mistake or some sort of collective madness—it would still be true that 7 is prime. Moreover, we are inclined to think that the truths of mathematics are independent of our very existence. Even before there were any human beings, or even if there had never been any human beings, it would still have been the case that $2 + 2 = 4$. For example, if there were 2 apples and 2 pears lying beneath a particular tree on a day exactly 20 million years ago, then it must still have been the case that there were 4 pieces of fruit under the tree, even though there were no intelligent agents in existence to recognize the fact.

It appears that the truths of mathematics are mind-independent, obtaining independently of facts about, or even the very existence of, the human mind. Yet such truths involve reference to individual abstract objects (the numbers), as we argued above. Then how could those objects, in turn, fail to *exist* independently of the human mind? Compare: it would still have been the case that Everest is the tallest mountain in the world, even if there had never been any human beings in existence to appreciate the fact. But this surely could not be so, unless Everest itself would still have existed in the absence of human beings. Mind-independent truth seems to require the mind-independent existence of the objects that figure in those truths. So, too, in the case of mathematics: if the truth of '$2 + 2 = 4$' is independent of the human mind, then the objects with which it deals (namely, the numbers 2 and 4) must have a mode of existence that is independent of the human mind.

It appears that there is a strong case for saying that abstract entities do have mind-independent existence. In this case our a priori knowledge of the properties of, and relations between, such entities (of the kind that platonists maintain is available to us through mathematical proof and conceptual analysis) will be genuinely substantive. But it does not yet follow that abstract entities have necessary existence, existing at all times in all possible worlds. There are of course many things that exist independently of our minds (Everest, for example) that do not have necessary existence. Are there, then, any arguments to support the stronger conclusion of necessary existence for numbers and universals?

There is indeed such an argument, premised upon the claim that the truths of mathematics, and many truths about universals, are *necessary*—being true with respect to all times in all possible worlds. Thus no matter how different the world might have been from the way it actually is, it would still have been the case that 2 + 2 = 4, and it would still have been the case that no surface can be red all over and green all over at once. We are strongly inclined to believe that truths such as these hold come what may, no matter what else may or may not be true of the world, and no matter how different the world might possibly have been. Then how could the numbers 2 and 4 not exist in all possible worlds, given that '2 + 2 = 4' is both true with respect to every possible world, and *about* the numbers 2 and 4? And since the truth 'No surface can be red all over and green all over at the same time' is about the universals redness and greenness (according to the arguments presented earlier), how could those entities, too, not exist in all possible worlds? It seems that truths that are necessary must require the necessary existence of the entities those truths concern.

The conclusion of this chapter is that there is a seemingly powerful case supporting platonism with respect to numbers and universals. There appear to be convincing arguments for believing that there are abstract entities that not only exist independently of the human mind, but that exist at all times in all possible worlds. In the next chapter we shall consider what an empiricist can do to undermine these arguments. We shall also consider whether an empiricist has any positive arguments for thinking that platonism is false.

3

The Empiricist Case against Platonism

HAVING set out the arguments supporting platonism in chapter 2, in this chapter I shall present the main arguments that, from an empiricist perspective, can be deployed against it.

Reducing Universals

The most urgent task facing an empiricist is to undermine the case for platonism. This has generally been attempted via a species of reductionism. The intention has been to show that what appears to be talk about abstract entities is really talk about something else, most usually internal relations between ideas in our minds. If one could thus reduce abstract-object talk to talk about non-contingent features of the human mind, it would have been shown that the truths of logic and mathematics, while a priori, are not genuinely substantive. For they would neither give us knowledge that is contingent (as is our knowledge of our own current mental states), nor knowledge that relates to any aspect of reality outside our minds.

To adopt this reductionist strategy is to try to show that the subject-matter of logic and mathematics can be interpreted adequately without reference to abstract entities. In connection with logic, and with the sort of conceptual analysis that one finds in philosophy, the task is relatively easy. We can interpret these disciplines as being concerned with relations between *rules*, conceiving of rules as mind-dependent entities, reducible in turn to the semantic intentions of those who employ those rules. Thus 'Anything red is coloured' may be understood to be true in virtue of a relation obtaining between the concepts *red* and *colour*, where these in turn are taken to be rules for classifying items in the world. And the rule of classification expressed by the word 'red' will itself consist in the manner in which most users of that word intend to continue to employ it. This is in fact the standard empiricist line: truths of logic and conceptual analysis are *analytic*, being concerned only with relations between concepts, where these in turn are understood to reduce to facts about the human mind.

(Note that while I have here identified concepts with rules of classification, most traditional empiricists have thought of them as *images*. But the weaknesses in the imagist theory of concepts have been long recognized. In fact no image, or sequence of images, can carry unaided the content of even a simple proposition such as 'Grass is green', let alone of a complex proposition like 'Life may be discovered on Mars in the next ten or twelve years'.)

Does any such form of reductionism about concepts have the resources to reply to the platonist arguments from objective truth and necessity? The natural move would be to interpret sentences that seem to involve generalizations about abstract universals as being, covertly, statements about concepts, or ideas in our minds. We should then regard the content of 'There is something that nothing is (namely a dragon)' as being given in reality by a sentence such as 'There is some idea that applies to nothing (namely the idea of a dragon)'. We cannot, however, take the most obvious option with such statements when they relate to situations in which there are no human beings. We cannot, for example, interpret 'Even if there had never been any human beings, it would still have been true that there is something that nothing is (namely a dragon)' to mean 'Even if there had never been any humans, there would still have been some idea that applies to nothing'. Nor can we interpret '20 million years ago there was something that grass and leaves had in common' to be saying '20 million years ago there was some idea that applied to both grass and leaves'. For these sentences would then imply the existence of ideas in situations in which there are no intelligent agents to possess them, which would be absurd.

In fact, however, there is no special problem here. We can rather take the first sentence to say that there *is* some idea (namely our—presently existing—idea of a dragon) that applies to nothing in a world that differs from ours only in never having contained any human beings. And we can take the second sentence to say that there *is* some idea that applies to grass 20 million years ago and to leaves 20 million years ago. All we need for this is the thesis that presently existing ideas can apply to things *across* time (as when we employ them in thoughts about the remote past or future), or *across* worlds (as when we employ them in thoughts about a world in which the human species never evolved), without having to exist *at* those times or *in* those worlds.

A similar move enables us to avoid the argument from necessary truth to the necessary existence of universals. For if this argument were to be valid, we should have to interpret the necessity of 'No surface can be red all over and green all over at once' as holding in virtue of a truth of the following sort:

> For all worlds, w, and all times, t, the concepts *red* and *green* exclude one another in w at t.

This would imply the existence of concepts (that is, universals) at the times and in the worlds where the original sentence is true (that is, at all times in all possible worlds). But in fact there is nothing to stop us regarding the necessity of that sentence as holding in virtue of a truth of the following form:

> For all ways of thinking of a world, *w*, and all ways of thinking of a time, *t*, the concepts *red in world w at time t* and *green in world w at time t* are (now) mutually exclusive.

The idea here is that no matter how we think of possible situations, our concepts (ideas) of red and green exclude one another, provided that they occur in thoughts about one and the same such situation. An empiricist can thus maintain that analytic truths, while being true in virtue of merely presently existing ideas, nevertheless constrain our talk and thought about all other times and possible worlds. We can thus hold on to the objectivity of at least some necessary truths (namely, those that are analytic), while being committed only to the existence of concepts as ideas in the human mind.

Notice, however, that in thus explaining away the kind of talk which lends credence to the existence of *abstract* universals, we are not thereby prevented from recognizing the existence of *immanent* universals, as forming part of the natural world. We can allow that there is a sense of 'Grass and leaves have something in common' that is not merely about ideas in our minds, but rather claims that there is something that is partially present *in* individual leaves and blades of grass, which explains their common appearance. I can see no reason why empiricists should object to the existence of immanent universals. For while such universals will be genuinely mind-independent, forming part of the natural universe just as much as individual mountains and trees do, we shall have no a priori knowledge of them. Nor will they exist in all possible worlds. On the contrary, it will be the business of the natural sciences to discover what common properties there really are in the actual world that explain the causal powers individual objects have in common with one another.

Reducing Numbers

Reductionism in mathematics is a more complex affair. This is partly because we seem to have terms here which purport to refer to individual things (that is, the individual numbers), and it is not so obvious how these may be reduced to rules or psychological phenomena. Nevertheless, many of those who pursued the logicist programme in mathematics around the turn of this century were attempting to show how the truths of arithmetic could be reduced to those of logic (with these in their turn being shown to be analytic). If this programme could be carried through, and if it could be shown that analytic truths are not about anything outside the human mind, then the mind-dependence of numbers, too, would have been established. Note, however, that it is possible to be a logicist and a platonist at the same time (as Frege in fact was). For one might claim that the truths of logic, and analytic truths generally, obtain in virtue of facts about necessarily existing universals. To reject platonism about numbers one might also need to reject platonism about universals (though not vice versa).

The first step in the logicist programme was to realize that statements of number of the form 'There are n Fs' are best understood as statements about concepts (or as about the sets of things those concepts pick out). In the last chapter we noted, in connection with the sentence 'Jupiter has 4 moons', that the number 4 is certainly not an attribute of each individual moon. It might be replied that this does not yet show that numbers are not ordinary attributes. For they may be distinctive in being attributes, not of individual things, but of *collections* of such things. But in fact there are severe difficulties with such a view. For one thing, the very same physical collection can have different numbers applied to it. Thus 1 pack of cards can also be described as 4 suits, or as 52 cards. Moreover, there are problems with the number 0. If the number of Mary's children is 0, then there exists no collection for 0 to be an attribute of.

These difficulties can easily be resolved if statements of number are statements about concepts. Then 'Jupiter has 4 moons' states of the concept *moon of Jupiter* that it is instantiated 4 times, and 'Mary has 0 children' says of the concept *child of Mary* that it is not instantiated at all.[1] In the same way, what changes when the pack of cards may variously be described as '1', '4', or '52' is the concept involved in the statement (*pack*, *suit*, and *card* respectively). But note that it does not follow from this that numbers themselves are attributes of concepts, or second-level concepts. For, as we noted in the last chapter, sentences of the form 'There are n Fs' can be rewritten as identities, saying, in effect, 'The number of things falling under the concept *F* is none other than n'. While the whole statement can be regarded as being about the concept *F*, the numeral 'n' may still be treated as a proper name.

The crucial consideration was then to notice that numerical identity can be defined as a relation between concepts. Thus to say that the number of Fs is none other than the number of Gs can be analysed as saying that the instances of the concept *F* may be placed in a one-to-one relation to the instances of the concept *G*, where the idea of such a relation can be defined without mentioning numbers. To get the feel of this suggestion, notice that a waiter can tell that the number of bowls on the table is the same as the number of plates without having to count, by seeing that every bowl stands on a plate, that every plate has a

---

[1] Alternatively, 'Jupiter has 4 moons' can be regarded as saying of the set of moons of Jupiter that it has 4 members, and 'Mary has 0 children' can be understood to say of the set of Mary's children that it is empty.

bowl standing on it, that all plates on which any given bowl stands are the very same (that is, that the bowl stands on just one plate), and that all bowls standing on any given plate are the very same (that is, that just one bowl stands on that plate).

The individual numbers can then be introduced recursively, defining 0 as the number of things that are not identical with themselves, 1 as the number of things that are identical with 0, 2 as the number of things that are identical with either 0 or 1, and so on.[2] In which case 'Mary has 0 children' may be analysed as saying that the concept *child of Mary* is one-to-one correlated with the concept *not identical with itself*. And to say that the Earth has 1 moon may be understood as saying that the concept *moon of the Earth* is one-to-one correlated with the concept *identical with 0*. And so on. Such an analysis need commit us to nothing besides concepts (reducible in turn to facts about the human mind, for an empiricist) and the relations between them.

The project of logicist reduction in mathematics went out of fashion in the early decades of this century, for a variety of technical reasons, though it has been revitalized again recently by Crispin Wright (contrary to his intention).[3] Other forms of reduction have also been attempted. Notable amongst these is the modal–structural approach of Geoffrey Hellman,[4] developing some ideas first suggested by Paul Benacerraf.[5] On this account, mathematics deals with the class of possible structures that share the properties of the natural number sequence (infinite extent, each member having a unique successor, and so on). The structures in question are (when instantiated) concrete ones—for example, the sequence of inscriptions '*', '**', '***', '****', and so on. So there is no commitment to abstract objects. Yet because the account is modal, dealing with all *possible* structures sharing certain properties, it can achieve the universality (particularly the applicability to other possible worlds) of mathematics.

---

[2] Frege's actual definitions were given in terms of sets of sets—as it turns out, disastrously, in the light of Russell's Paradox. See Crispin Wright, *Frege's Conception of Numbers as Objects* (Aberdeen: Aberdeen University Press, 1983), chs. 3 and 4, for further discussion.

[3] See his *Frege's Conception*, ch. 4.

[4] See his *Mathematics without Numbers* (Oxford: Oxford University Press, 1989).

[5] See 'What Numbers Could Not Be', *Philosophical Review*, 1965.

It remains true, however, that it would be a fearsome achievement to carry through to completion any version of the reductionist project. It might therefore be wise for empiricists to adopt some alternative strategy for opposing platonism in mathematics if they can, which will carry less risk of failure on technical grounds. For if there is a lesson to be learned here from the history of philosophy, it is that conceptual reductions are remarkably difficult to complete successfully. This is especially likely to be the case where our discourse is very highly developed and complex, as is that of mathematics. (Compare the failure of the phenomenalist reductive programme, which attempts to reduce talk of physical objects to talk about patterns in experience.)

Mind-Dependent Existence

There is, however, another strategy available to an empiricist, which carries less risk of failure. We can grant that mathematics is concerned with abstract objects, making no attempt to analyse away such apparent references to numbers as occurs in '7 is prime'. But we can deny that these objects have an existence that is independent of the human mind. We can rather claim that their existence supervenes upon aspects of human activity —particularly on the rules implicit in the practices of those who count, add, and subtract—in such a way that they would not have existed if human beings (or other intelligent agents) had never existed. So we may allow that there are abstract objects, but deny that these objects exist necessarily, or indeed independently of the human mind. If fact we can adopt for numbers just the sort of position that we sketched concerning the mode of existence of sentences in the last chapter.[6]

But if numbers have only mind-dependent existence, how are we to explain the fact that we apply arithmetic to remote regions of time, for example in calculating the orbit of the Earth many millions of years in the past or future (when, we may presume, there are no minds for numbers to supervene *upon*)? Indeed, what of the point that even if there had never been any human

[6] A view of this sort is defended by Karl Popper in *Objective Knowledge* (Oxford: Oxford University Press, 1972).

beings, it would still have been true that 2 apples and 2 pears would constitute 4 pieces of fruit? How can this be so, if in such a case there would have been no such things as 2 and 4? It is worth exploring in some detail the way in which an empiricist might respond to these problems. The general strategy will be as it was in the discussion of universals above, namely to try to account for the truths in question in terms of the *present* existence of numbers alone.

How is one who believes that numbers do genuinely exist as abstract objects, but mind-dependently, to account for a truth such as the following?

> Even if there had been no intelligent agents, the number of moons of Jupiter would still have been 4.

Suppose we let 'a' designate the actual world. Then we might try expressing the above truth as follows:

> For all possible worlds w, if w differs from **a** only in containing no intelligent agents, then the number of moons of Jupiter (in w) is none other than 4 (in **a**).[7]

The thought here is that we can avoid commitment to the existence of the number 4 in any other world besides the actual. The trouble, however, is that the number 4, in **a**, surely cannot be identical to (be none other than) the number of moons of Jupiter in w without itself *existing in* w. What we really require is something having the following form:

> For all possible worlds w, if w differs from **a** only in containing no intelligent agents, then the number (in **a**) of moons of Jupiter (in w) is none other than 4 (in **a**).

Here all references to numbers are confined to the actual world. The problem is to find a form of account that would have such a feature. For it is by no means easy to see how a single definite description (in this case 'the number of moons of Jupiter') can contain references to two distinct possible worlds at once.

I take it that all platonists should agree that statements of the form 'There are n Fs' are really statements of numerical identity, having the form 'The number of Fs is none other than n'. They

---

[7] I should stress that talk of 'possible worlds', for an empiricist, is merely a convenient mode of speech, carrying no commitment to the real existence of worlds other than our own.

should also accept that the criterion of numerical identity is one-to-one correlation between concepts, in the manner explained earlier. Then our target statement may be expressed by something having the following general form:

> For all possible worlds w, if w differs from **a** only in containing no intelligent agents, then the instances of the concept *moon of Jupiter* (in w) one-to-one correlate the number of things such that . . . (in **a**).

Here the dots would be filled in by the appropriate definition of the number 4. In such a case we would be relying upon relations of one-to-one correspondence *across possible worlds* (that is, between w and **a**) to avoid commitment to the existence of numbers in any but the actual world.[8]

But what of these relations of one-to-one correspondence? How can they both obtain across possible worlds and be mind-dependent? The answer is that they, in their turn, are not actual relations but possible ones. For consider what it must mean to say that there are n Fs, if an empiricist insists that concepts (and relations) are best understood as rules of classification. Certainly not that there is any actual rule of correspondence relating the number n to the instances of the concept *F*. For in many cases no such rule will have been set up. Rather, it must mean that it is *possible* to create such a rule, for example by counting. In full, then, our account of our target statement would run as follows:

> For all possible worlds w, if w differs from **a** only in containing no intelligent agents, then there is another possible world v which differs from w only in whatever is required for there to be both numbers and a rule of correlation such that the instances of the concept *moon of Jupiter* (in v) one-to-one correlate the number of things such that . . . (in v).[9]

---

[8] Even if platonists do not accept the need for definitions of the individual numbers, they must certainly agree that descriptions of the form 'The number of Fs' will come to designate a particular number in virtue of some relation obtaining between the number in question and the instances of the concept *F*.

[9] Of course there will need to be restrictions in applying this scheme, so that if the intelligent agents in v are to be counting intelligent agents, then they must count all *with the exception of themselves*. Otherwise the sentence 'Even if there had been no intelligent agents, the number of intelligent agents would still have been 0' will come out as false rather than true.

While this does not quite restrict all references to numbers to the actual world, it does in effect restrict them to worlds in which there are intelligent agents, which is all that we need for numbers to have mind-dependent existence.

We can deal in similar manner with the necessary status (truth about all possible worlds) of mathematical statements such as '2 + 2 = 4'. What initially causes us a problem here is that such truths seem naturally to be represented in the following form:

> For all possible worlds w, it is a truth about w that 2 + 2 = 4.

For it is hard to see how a statement about numbers can express a truth *about* a world, without numbers themselves existing *in* that world. But in fact we can express the content of our new target statement somewhat as follows:

> It is a truth about **a** that 2 + 2 = 4, and for all possible worlds w, the truth (in **a**) of 2 + 2 = 4 is *applicable to* w.

Here the idea of the applicability of a numerical truth is to be cashed out in a similar manner to that outlined above, involving relations across worlds between numbers and the instances of concepts. Thus the applicability of 2 + 2 = 4 to a world in which there are no intelligent agents would consist in the following sorts of facts. If the instances of the concept *apple under the tree* in that world are one-to-one correlated to the number 2 in **a**, and if the instances of the concept *pear under the tree* in that world are similarly correlated to the number 2 in **a**, then the instances of the concept *apple or pear under the tree* in that world are one-to-one correlated to the number 4 in **a**. Here we have preserved the intuitive thought that numerical equalities are necessary (that is, applicable to all worlds) without commitment to the existence of numbers in any worlds but those in which there are intelligent agents.

Can there be Causal Contact?

It appears that an empiricist can adequately undermine the arguments that seem to support platonism. But does an empiricist have any direct objection to platonism? We have seen that there is insufficient reason for believing platonism to be true, but are there any good reasons for positively believing it to be false? The main argument to this effect arises out of the question whether a platonist can give an adequate account of our supposed knowledge of necessarily existing abstract entities.

As we noted in the last chapter, many platonists have believed that we possess a special faculty of intellectual intuition, modelled by analogy with sense-perception. However, it is doubtful whether this quasi-perceptual model is really coherent. For how are we to make sense of the idea that a changeless, necessarily existing entity (which is what platonists suppose the number 7 to be) can have causal powers? The problem here is that the concept of causality is at least closely bound up with counterfactual and subjunctive conditionals. (A counterfactual conditional has the form 'If X *had* been the case, then Y *would have* been the case'. A subjunctive conditional has the form 'If X *were* to be the case, then Y *would be* the case'.) To say that A caused B is to imply that if A had not been the case, then B would not have; and also to imply that if in other circumstances something sufficiently similar to A were to be the case, then something similar to B would be also.

Thus, if it is true that a spark caused a particular explosion, then it must be true that if the spark had not occurred, nor would the explosion (other things being equal). It must also be true that in all other sufficiently similar circumstances a spark would be followed by an explosion. But where A is something that is necessarily the case, the supposition 'If A had not been so' will make no sense. If the fact that 7 is prime obtains at all times in all possible worlds, as platonists believe, then it will be impossible to suppose that 7 had not been prime. In which case it cannot be the fact that 7 is prime that causes our belief that it is; indeed such a fact cannot be a cause at all.[10]

Although the quasi-perceptual model is incoherent, it is not the only possible way in which platonists might conceive of causal access to the abstract realm. An alternative model would be a gravitational one. Just as it is the fact that an asteroid enters the gravitational field of the Earth rather than of Mars (which may for these purposes be thought of as changeless) that

---

[10] Plainly, to suppose that the existence of numbers is timeless rather than necessary would not help at all here. For how could an atemporal entity be the cause of a temporal event, such as the onset of a belief?

explains its subsequent orbit, so it may be the fact that one intuits the nature of 7 rather than of some other number that explains our subsequent belief that 7 is prime. We may thus think of abstract objects as having associated with them permanent (indeed necessarily existing) 'intuition-fields', such that when the mind stands in the intuition-relation to one abstract object rather than another, it is caused to have one belief rather than another. This gives us appropriate counter-factual and subjunctive conditionals. We may truly say that if someone had not intuited the nature of 7 at that moment, then they would not have come to believe that 7 is prime; and if in other similar circumstances they were to intuit the nature of 7, they would then believe that 7 is prime.

This account enables us to make some headway in explaining how platonic entities can be causes. But it, too, faces severe problems. For even if abstract objects are associated with permanent intuition-fields, it still has to be possible to explain why any given object is associated with one such field rather than another. Yet such explanations will return us to the sort of unintelligible counterfactuals met with earlier. Consider the gravitational case again. Even if it is the bare fact that an asteroid enters the gravitational field of the Earth rather than of Mars that explains its subsequent pattern of movement, it must surely be possible to explain *why* the Earth has the gravitational field that it has. And indeed we can provide just such an explanation, in terms of the Earth's mass.

The general moral is that relational causal powers need to be explicable in terms of the categorical properties of the things that possess those powers. That is to say, if something has the power to affect in a given way things that come to stand in a certain relation to it, then this must ultimately be explicable in terms of the non-relational properties that object possesses. So if the number 7 has the power to cause those who intuit it to believe that 7 is prime, this must be causally explicable in terms of the properties of the number 7. But this will commit us once again to statements such as 'If 7 had not been prime it would not have had the powers that it has'—which are, as we saw above, unintelligible (given that 7 has necessary existence, and possesses its attributes necessarily).

A further problem with the idea that platonic objects can have effects on the human mind (in either of the above versions), is that it requires us to recognize a whole new species of causality, hitherto unknown to science. For we would have to believe that an abstract object (non-physical and changeless) can have effects within the natural (physical and changing) world. The difficulty faced here by platonists is similar to that faced by dualists in the philosophy of mind. Those who believe that the mind is a non-physical entity, or who believe that mental events such as thoughts, decisions, and intentions are non-physical ones, are then faced with the problem of explaining how there can be causal contact between non-physical events and physical ones. For we do normally suppose that our thoughts and decisions *cause* our bodies to move in certain ways. Now I doubt whether there is any objection in principle to the idea of such causation. But it does conflict with a well-established working hypothesis of science, namely that all causes are physical ones.

Science has progressed by ignoring the possibility of causation by spirits or other non-physical entities, and by looking wherever possible for physical mechanisms. So the immense explanatory success of science gives us reason to be doubtful, at least, about both the quasi-perceptual and the gravitational models employed by platonists to explain our knowledge of logic and mathematics. If we are impressed by the success of science, we shall wish to insist that such knowledge should be explained *naturalistically*, in terms of the operation of laws and processes that science gives us some reason to believe in. We shall return to consider this idea in some detail in chapter 9.

Platonism and Scepticism

Yet another objection to platonism is that it seems inevitably to lead to scepticism with respect to our knowledge of logic and mathematics. For if platonism were true, what reason could we have for thinking that intellectual intuition is generally reliable? It might be claimed that we have just the same reasons here as we have for thinking that sense-perception is reliable, namely that the hypothesis of reliability provides the best overall explanation of the existence and coherence of our beliefs. But in fact our belief in perceptual reliability fits into an explanatory network that is wholly lacking in the case of intuition of abstract objects. For we have beliefs about our perceptual apparatus itself, and its mode of operation, that mesh closely with our beliefs about the physical world that we perceive. And these in turn provide the best available explanation of the course of our experience. (These ideas will be developed more fully in chapters 11 and 12.)

In contrast, we have no beliefs about the structure of our faculty of intuition or its mode of operation. Indeed, the existence of such a faculty remains a bare hypothesis. Nor can the hypothesis of its reliability play any real explanatory role. In particular, we cannot use it to explain our success in building bridges and aeroplanes (in which mathematics plays a significant part), unless we can suppose that the way things are in the abstract realm can make a difference to what happens in the physical realm—which brings us back to the incoherent idea of abstract objects as causes once again. In fact all we need to suppose is that mathematical calculation gives us access to whatever structural features of the physical world underlie our success in action.

The only other option available to a platonist, in explaining our knowledge of the abstract realm, is to claim that such knowledge is innate, being brought to consciousness by the activity of thinking and reasoning. There are two subtly different versions of such an account. Either it can be claimed that beliefs about the abstract realm are already latent in us prior to thought and reasoning. Or it can be claimed that the faculty of reason itself has a constituent structure, which is such that its operation gives rise to beliefs about the abstract realm. But the differences between these two forms of account need not detain us at this stage. For in either case the idea is that there is something innate in the human mind that mirrors or depicts the nature of the abstract realm.

Such an account enables a platonist to do without the supposition of causal contact between an abstract object and the human mind, and this is a mark in its favour. But we are still left with the same problem of explaining what reason we have for supposing that the mirroring or depicting is generally accurate. For suppose we allow that we have innate beliefs about logic and mathematics. If we grant the platonic conception of the subject-matter of these disciplines, then what reason would we have for thinking that the beliefs in question are generally true?

Since we have no idea how it is that these beliefs might come to be innate, nor any account of why the process that leads them to be innate should ensure that they mirror the nature of the abstract realm, we lose nothing in explanatory power if we suppose that our logical and mathematical beliefs are generally false. (We shall return to this point in greater detail in chapter 10.) For again, we cannot appeal to the practical success of our applications of logic and mathematics as a reason for thinking that our innate beliefs are reliable, unless we can also explain how truth about the abstract realm could make a difference to the course of events in the physical world. But this we are debarred from doing, given that the idea of causal influence of the abstract on the physical is incoherent.

It would appear, then, that neither the hypothesis of a faculty of intuition nor the hypothesis of innateness are capable of explaining how we have knowledge of the abstract realm (given the platonist's conception of it). We then face a choice of either denying that we may have reasonable beliefs about logic and mathematics, or of finding some alternative (non-platonic) account of their subject-matter.

In conclusion: there are insufficient reasons for supposing platonism to be true. Yet there are very good reasons for supposing it to be false. For our acceptance of platonism would mean both that we could give no account of the manner in which we might obtain knowledge of the abstract realm and that we could no longer hold reasonable beliefs about logic and mathematics, which would purportedly concern such a realm. The empiricist case against this form of substantive a priori knowledge (arguably the most plausible form, as we saw at the beginning of the last chapter) is therefore a powerful one.

4

The Empiricist Case against Nativism

Few empiricists besides Locke have bothered to present explicit arguments against nativism, mostly taking its falsity for granted. In the present chapter I shall consider whether Locke's case is really powerful enough to justify such an attitude.

Locke on Innate Knowledge

Locke's main argument against the existence of innate knowledge in Book I of the *Essay* is that the various supposed innate truths (for example, 'God exists' and 'Whatever is, is') are not in fact universally assented to. He cites the examples of children and madmen, many of whom will not assent to such propositions if they are put to them. Locke is therefore assuming that innate knowledge would, if it existed, necessarily have to be present in everyone, and also that it would have to be available to consciousness from birth. As we shall see, both of these assumptions are false.

The argument from madmen is certainly a bad one. To say that something is innate for human beings is to say that all normal members of the species will possess it, not that all without exception will do so. Consider, for example, the fact of possessing ten toes. This is surely an innate feature of human beings. But some humans will in fact have less, and some more. Some may lose a toe in an accident, and occasionally babies are born with an extra toe, or with a toe missing. So the fact that madmen (who on any account of the matter are not normal human beings) lack some knowledge that the rest of us possess does not show that such knowledge is not innate.

The argument from children is also unsound, but needs to be handled somewhat differently. A natural first response to it would be to claim that innate knowledge might be *latent* in children—that is to say: it is there, but not yet available to consciousness. Locke anticipates this reply, and responds by adopting what might be called 'the principle of mental transparency'. He claims that there cannot be anything in the mind that the subject is unaware of. But this principle is surely indefensible. Nor need we commit the anachronism of appealing to Freudian theories of the unconscious to show as much. For consider the everyday phenomenon of temporary memory loss. You may know that you know your mother's birthday, but be unable for the moment to recall it. Then an hour later you may be able to remember it again. If we accepted the principle of mental transparency, we should have to say that you started by having the knowledge of your mother's birthday, then you lost it, and then you acquired it again without any process of learning. This is surely absurd. Rather, we should say that the knowledge was *in* you throughout, but that for a short period it was not accessible to consciousness.

While Locke's explicit argument against latent innate knowledge is inadequate, it may seem that he has a valid point nevertheless. For are we really prepared to accept that a newborn infant has its head already stocked with a range of actual (if as yet merely latent) knowledge? One way of developing this point is to notice that you cannot have knowledge of something unless you also believe it. (You cannot *know* that the Earth is getting warmer unless you at least *believe* that the Earth is getting warmer.) And what makes a mental state a state of belief is that it is apt to interact with your other mental states (particularly desires and intentions) in such a way as to control behaviour. Thus, what makes the difference between *believing* that the Earth is getting warmer and *hoping* that it is, for example, is that you are prepared if necessary to act on it. You may consider moving to Iceland, or stop using aerosols (depending upon your other beliefs, and on what it is that you want). Then since an infant cannot manifest any of the appropriate behaviour, it would seem that it cannot have articulate beliefs either.

This argument is perhaps overly swift, however. For even if an infant cannot be said to be born already possessing beliefs, it may be that it is born with a stored stock of *propositions*. These may start by being inert, but then become knowledge as soon as the child is old enough to be conscious of them. Yet even this may strike one as an extraordinary hypothesis. The idea that the head of an infant is already stocked with a range of articulate propositions may strike one (and does strike me) as just *wild*. At any rate, if this were the only form that nativism could take, then the burden of proof would surely fall squarely on the nativist to provide some convincing argument for their view. Locke would be quite right that there is a strong presumption against nativism, unless and until we are shown otherwise.

In fact, however, there is quite a different sense in which innate knowledge may be latent from birth. For it may be innately determined that children develop such knowledge at a certain stage in the course of their normal growth, irrespective of details of education and experience. Compare, for example, the possession of pubic hair. This is surely an innate feature of adult human beings. But it is not present from birth, only making its appearance with the onset of puberty. Similarly, then, in the case of knowledge: it may be that it is not present at all at birth (there being no stored stock of propositions), but that it is innately determined that such knowledge will make its appearance at some particular stage in normal cognitive development.

Locke himself considers ideas related to this one. For he argues against the suggestion that truths may be innate in the sense that a child has an innate *capacity* for knowing them. He also argues against the thesis that innate knowledge may make its appearance in the mind when the subject first attains the use of reason. But his only response to the first suggestion is that one cannot then distinguish between truths that are learned and truths that are innate, if 'innate' just means that a child has an innate capacity to know them. While this may be true, if 'capacity' is understood broadly, it is not an objection to the developmental thesis sketched above. For if the truths that make their appearance subsequent upon experience *could not have been learned from* that experience, then this will be sufficient reason to count them as innate, as we shall see in the next section. And as for the suggestion that innate knowledge may make its appearance with the onset of reason, Locke's response is mostly to chip away at this as a proposed *time* at which innate knowledge might appear, which in no way touches the general idea behind developmental nativism.

Varieties of Innateness

We have noted that while one form of nativism claims (somewhat implausibly) that knowledge is innate in the sense of being present as such (or at least present in propositional form) from birth, it might also be maintained that knowledge is innate in the sense of being innately determined to make its appearance at some stage in childhood. This latter thesis is surely the most plausible version of nativism. Indeed, there seems no particular reason why we should presume its falsehood. For given that much of the physical growth and development of human beings is innately determined, why should the same not be true of our cognitive development also? It is therefore not at all obvious that the burden of proof is on the defender of this form of nativism to make out their case, rather than on Locke to show that all knowledge is in fact acquired from experience. But we should now notice that this second hypothesis, in turn, admits of two alternative versions.

First, a belief might be innate in the sense that it is acquired in any course of experience sufficient for forming beliefs at all. To adopt this hypothesis is to allow (as seems likely) that an infant sensorily deprived from birth would never have any knowledge or beliefs, innate or otherwise. We would be allowing that some initial experience is necessary for the mind to operate normally, and for innate knowledge to make its appearance. But we would be claiming that it does not matter *what* experiences the child has, provided that they are sufficiently rich and varied for it to acquire at least some beliefs. Let us call this 'the hypothesis of general triggering of innate knowledge'—'general triggering' because almost any experience will serve, the content of the experiences needing to bear no relation whatever to the content of the innate beliefs that make their appearance as a result.

The second sense in which an acquired belief might be innate would be if its existence were inexplicable on any model of learning, its content being such that it could not have been learned from the experiences that gave rise to it. To adopt this hypothesis would be to allow, as before, that a sensorily deprived infant would never come to have any beliefs. But it would also be to allow that quite specific types of experience may be necessary for a given innate belief to make its appearance. For example, it might be claimed (as we shall see in chapter 6) that our knowledge of grammatical structure is innate, but that some experience of language is necessary to trigger this knowledge into existence. Or it might be claimed (as we shall see in chapter 8) that our knowledge of the rudiments of human psychology is innate, but that some experience of other humans is necessary for it to make its appearance. (So Tarzan, brought up by apes in the jungle, would never acquire either of these sorts of belief, although his experience would be sufficiently rich for him to have many other beliefs.) But the knowledge in question could still reasonably be said to be innate, provided that it is impossible to see how it could have been *learned* on the basis of the experiences in question—for example, if no combination of memory, induction, and inference to the best explanation could have generated that knowledge from such a meagre basis.

Let us call this second version of developmental nativism 'the hypothesis of local triggering of innate knowledge'—'local' because specific (content-relevant) types of experience are necessary for the knowledge to make its appearance, but still 'triggering' because the content of the knowledge acquired is so related to the content of the experiences that give rise to it that the former could not have been learned from the latter. Most of the arguments in support of nativism that we shall consider in later chapters are in fact arguments for local triggering. Notice that Locke himself provides no direct arguments against either of these forms of developmental nativism.

Locke on Concept Acquisition

The arguments considered above from Book I of the *Essay* are not the only ones that Locke uses against nativism. Indeed, they do not even constitute his main argument. Rather, he thinks that it will be sufficient to refute nativism if it can be shown that the hypothesis of innate knowledge is an *unnecessary* one—that is to say, if he can provide an alternative account of the genesis of all knowledge in experience. This is his strategy throughout the remaining three books of the *Essay*.

Now, although I denied above that there is a general presumption against the truth of nativism (at least in either of its more plausible developmental versions), it does seem to me that Locke is on strong ground here. For suppose that both Locke and the nativist could provide equally good explanations of the knowledge we actually possess. In the case of the nativist (but not of Locke) there would still be something left over that needed explaining—namely, how it is that some of our knowledge comes to be innate. We already know that some knowledge is derived from experience, and we know roughly how this takes place (through perception). So the hypothesis that all knowledge comes from experience leaves nothing further in need of explanation. In contrast, if some knowledge is innate, it still remains to be explained how it comes to be so. Therefore, other things being equal, Locke's hypothesis is the one to be preferred, since it leaves less in need of explanation.

In fact Locke does not focus very directly on explaining the acquisition of knowledge from experience. Rather, most of his efforts are directed towards showing how all our concepts (ideas) may be derived from experience. He here assumes, I think, that no knowledge can be innate if no concepts are.[1] So if he can show that the hypothesis of innate *concepts* is an unnecessary one, he believes that he will thereby have shown that the hypothesis of innate knowledge is also unnecessary.

The idea that innate knowledge requires innate concepts is certainly a very plausible one. For it is clearly the case that knowledge itself requires concepts. Knowledge (at least in the sense that concerns us) is essentially propositional—it is always knowledge *that such-and-such is the case*. So you cannot possess such knowledge unless you also possess the concepts involved in the proposition *that such-and-such*. You cannot know that grass is green unless you possess the concepts *grass* and *green*. But in fact, whether this dependence of knowledge upon

---

[1] Indeed, see the *Essay*, bk. i, ch. iv, sect. 20.

concepts extends also to the dependence of *innate* knowledge upon *innate* concepts turns on what exact concept of the innate is in question.

Clearly there can be no innate knowledge in the sense of stored propositions unless there are also innate concepts. For the constituent concepts of those propositions will also have to be present in the mind from birth if the propositions themselves are. Nor can there be general triggering of innate knowledge by experience unless there is also general triggering of innate concepts. For remember that the experiences that give rise to such knowledge need bear no relation to it in content, and so would not be the sort of experience from which one could derive (that is, learn) the concepts in question either. Matters are quite different, however, when it comes to the thesis of local triggering. For, on this account, it may be that concepts are learned from appropriate and relevant experience, but are then triggered into items of knowledge that go far beyond the content of the experiences that gave rise to the constituent concepts. It may thus be that while no concepts are innate, some knowledge is. For it may be that while the data that gives rise to our knowledge of some subject-matter is not sufficient to sustain the view that we acquired that knowledge through *learning*, it may still be sufficient to support the view that we learned the constituent concepts.

Thus Locke's assumption that there can be no innate knowledge without innate concepts is false. In which case his arguments against concept-nativism will not necessarily undermine knowledge-nativism. Nevertheless, it is worth considering his theory of concept acquisition in its own right. For, after all, most nativists have in fact held, not only that some knowledge is innate, but also that some concepts are. Moreover, the most plausible case of innate knowledge, to be defended in chapter 8 (namely, knowledge of the general principles governing our own and other people's psychology), will in fact be such as to involve the claim that there are innate concepts also.

Locke believes that we derive simple concepts from experience by *abstraction* (complex concepts can then be formed from simple ones by definition). The idea is that from a sequence of experiences we are to isolate the various features they have in common. We are to do this by ignoring differences of time, context, and so on, and by noticing and isolating recurring aspects. Consider this analogy. Suppose that you are taking part in one of those psychological experiments where you are handed a sequence of cards on which geometrical shapes of varying colours have been printed. Some are red triangles, some green triangles; some are blue circles, some red circles; and so on. As you are handed each card the experimenter says either 'This is a grink' or 'This is not a grink', your task being to acquire the concept *grink*. What you would do, of course, would be to attempt to spot resemblances between those cards that contain a grink, using those that do not to confirm or refute your hypotheses. This will be how Locke thinks of concept acquisition in general. For what you would in effect be doing in this experiment is abstracting from the sequence of your experiences the common feature of all grinks.

As it stands, however, Locke's account of concept acquisition appears viciously circular. For noticing or attending to a common feature of various things presupposes that you already possess the concept of the feature in question. Thus in order to notice that Peter, Paul, and Mary have something in common— namely, that they are all freckled—you must already possess the concept of being freckled. (This is not to say that you must already have a *word* for 'freckled'; but you must know in general how to distinguish people who are freckled from people who are not.) Even more obviously, if one thinks of concept acquisition as being a matter of formulating and testing hypotheses (for example, 'Is a grink any four-sided figure that is either red or green?'), then some concepts must already be possessed in advance (namely, the concepts that figure in the hypotheses).

While Locke discusses abstraction in terms that suggest that the processes involved will be conscious ones, it would be open to him to respond to the above objection by denying this. He could say that his theory is really that prior to acquiring any concepts there are processes that are *somewhat like* those of noticing resemblances, ignoring differences of context, and so on, only that they are non-conscious ones. This saves the account from circularity. But notice that it now apparently concedes that there are innate concepts after all, since non-conscious noticing of a resemblance presumably requires a non-conscious concept of the feature in question. So it is doubtful whether Locke would find this defence of his position satisfying. But how else is he to defend it? While he remains wedded to the language of abstraction, it appears that he must be committed to the mind's possession of conscious or non-conscious concepts prior to the process of abstracting a concept from experience.

Complex Concepts

If we set aside the worry about circularity outlined above, then Locke's theory can seem quite powerful. In particular, it has the resources to rebut many of the arguments presented by rationalists in favour of the thesis that there are innate concepts. In the *Phaedo*, for example, Plato argues that the concept *straight* must be innate because, first, judgements of 'almost straight' presuppose prior grasp of the concept *straight*, and, secondly, because no object in the world of our experience can be better than almost straight. Locke can reply by denying the implicit assumption that *straight* is a simple idea, which would have to be learned directly from experience if it is learned at all. Rather, he can say that what we derive from experience is the comparative concept *straighter than*, by observing pairs of unequally straight things. We can then introduce the concept *straight* by definition, as a thing straighter than anything else could be.

Locke can reply similarly to Descartes's argument in his third *Meditation*, that our concepts of God's perfections must be innate, since they plainly could not have been acquired on the basis of experience. Locke can maintain that what we acquire from experience are the various comparative concepts *better than*, *more powerful than*, *more knowledgeable than*, and so on. The idea of God may then be introduced by definition, as the one and only person who is better, more powerful, and more knowledgeable than anything else could possibly be.

However, Locke's account of concept acquisition does face a number of further difficulties. The most notorious is that there are many concepts that cannot be abstracted directly from experience, and yet appear not to be definable in terms that can be so abstracted either. Consider, for example, the concept of causation, which was treated at length by Hume.[2] All that is immediately observable in a case of A causing B is that the one precedes the other. Hume suggests that the remainder of our concept is made up from observing regular concurrences of such events. Then to say that A caused B will be to say that it preceded it, and that all events of the same type as A precede events of the same type as B. Yet clearly this falls short of our intuitive concept of a cause, which includes the idea that causes somehow necessitate their effects. Hume has to conclude that this idea is an illusion, brought about by our own habits of mind in expecting an event of type B when we see one of type A. Since our psychology is such that we cannot avoid expecting B whenever we see A occur, Hume suggests that we mistakenly assume that it is B itself that is unavoidable, given that A has occurred. He claims that what we do is to project a feeling of psychological necessity on to the world.

There are many reasons for rejecting Hume's account of the matter. One is that it tacitly assumes that we do have a concept of causal necessitation, if only as holding between psychological events. For what, otherwise, is the feeling of necessity a feeling *of*? Another point is that one may come to believe in a causal relationship between events after observing just a single instance, when there has been no opportunity for a habit of expectation to be formed. For example, I may see someone fall downstairs and come to believe that this caused their broken leg, although this is the first occasion on which I have observed anything of the sort to occur. Another argument is that we may *wonder whether* A caused B, where this is clearly more than a matter of wondering whether all events of type A precede events of type B. Yet in such cases we are obviously not wondering whether our minds are necessitated to expect B given an observation of A.

In fact what is really involved in our idea of causation are, at least, counterfactual and subjunctive conditionals, as we have already noted in chapter 3. To say that A caused B is to imply that if A *had not* happened, then B *would not have*.[3] It is also to

---

[2] See the *Enquiry Concerning Human Understanding*, sects. III-VII, and the *Treatise*, bk. I, pt. III.

[3] There is just one passage in his writings where Hume concedes this, without apparently noticing what it is that he has done. Thus in the *Enquiry*

imply that if in other sufficiently similar circumstances an event of type A *were to* happen, then an event of type B *would* happen also. Yet it is impossible to see how the concepts of such conditionals may either be abstracted from experience, or defined purely in terms that may be so abstracted.

Indeed, it is arguable that there is more to the concept of cause even than this. David Armstrong, for example, maintains that causation is best understood as a relation of necessitation between immanent universals.[4] This relation is held to imply, and hence to explain, the truth of the counterfactual and subjunctive conditionals mentioned above, rather than being constituted by them. Then it is *because* A is made up of properties that necessitate the properties making up B, that it is true that if A had not happened B would not have, and also true that if an event similar to A were to happen, so would an event similar to B. Moreover, Armstrong argues that the notion of causal necessitation has to be taken as primitive (that is, as indefinable). He also concedes that it cannot be acquired from experience, without, I think, noticing that he is therefore committed—very plausibly, in my view—to the claim that the concept of cause is innate.

Other concepts besides causation have led to problems for classical empiricists. Thus both Berkeley and Hume drew the conclusion that our concept of mind-independent, continuously existing physical objects is illusory, because they were (rightly) unable to see how such a concept could be derived from experience, or defined in terms of concepts that can be so derived. Since one cannot have experience of an unexperienced object, it is hard to see how one could derive the concept of such an object from experience. But then neither can that concept be defined in other terms derivable from experience. The closest we could get would be to say that an unexperienced object is the continuously existing *cause of* our episodic experiences. But this is plainly too broad (quite apart from the problem of how we are supposed to have acquired the concept of cause). It cannot

Concerning Human Understanding, sect. vii, pt.ii, he rephrases his definition of causality by saying 'if the first object had not been, the second had never existed'.

[4] See *What is a Law of Nature?*, ch. 6.

distinguish between the chair, as cause of my experience of it, and Descartes's all-powerful demon.

It is worth remarking here just how powerfully these empiricists must have been convinced of the thesis that there are no innate concepts. For rather than give up this thesis, they were prepared to deny that it is possible for us to conceive of a mind-independent physical reality. But in the absence of any convincing argument in its support, the proper conclusion to draw is surely that it is their anti-nativism itself that is false.

Simple Concepts

In fact problems arise for empiricists even in connection with the very simplest concepts, such as those of colour. For it is false that all instances of a given colour share some common feature. In which case we cannot acquire the concept of that colour by abstracting the common feature of our experience. Thus consider the concept *red*. Do all shades of red have something in common? If so, what? It is surely false that individual shades of red consist, as it were, of two distinguishable elements: a general redness together with a particular shade. Rather, redness consists in a continuous *range* of shades, each of which is only just distinguishable from its neighbours. Acquiring the concept *red* is a matter of learning the extent of the range.

Nor will it help to say, as Berkeley does,[5] that the concept of red is not an idea of a common feature abstracted from differing red things, but is rather an idea of a particular shade of red that is then used as a representative of the whole range. For there is nothing in the particular shade itself that can give you the extent of the range. Nor can this be learned from experience. There is nothing *in experience* that can tell you where in the spectrum red begins and ends. It would seem that the boundaries between the various colours must somehow be specified innately, unless they can be explained as taught social constructs of some kind.

This last remark suggests an alternative strategy that is available to empiricists for explaining our possession of concepts, and it is worth considering why they have not, in general, been inclined to pursue it. The strategy would be to appeal, not

[5] See his Introduction to *The Principles of Human Knowledge*.

to abstraction, but to some sort of linguistic training. Why should empiricists not say, consistently with their anti-nativism, that we are *taught* to classify things in the way that we do? On this view, possession of concepts would still arise out of experience by a process of learning. But the experience in question would not be (or not primarily) of the things to which the concepts apply, so much as of the norms that are prevalent in the person's language community. A child's first fumbling use of words would gradually be refined and perfected through a process of reward and correction. It would be, for example, by mistakenly describing an unripe tomato as 'red', and being put right by its parents, that a child would acquire its grasp of the boundaries between the colours.

However, the obvious question arising for such an account would be this: from where did our teachers, in turn, acquire their concepts? The answer, in terms of the theory, is equally obvious: from *their* teachers. But now we have a problem. Plainly the sequence of past teachers cannot be infinite, since the human race has not always been in existence. So it appears that there must have been some person, or group of people, who were the first to use simple concepts, without having been taught to do so. But then we shall be landed back with some version of abstractionism again, if we are to avoid commitment to nativism. For those first users of concepts will somehow have to have acquired their concepts directly on the basis of their experience. Certainly the problem of concept acquisition cannot be solved merely by pushing it back into the past. It is for this reason (among others)[6] that most empiricists have not taken very seriously the idea that we acquire concepts through linguistic training.

We should, however, beware of the suggestion that there must have been a first concept-user, if present concept-users get their concepts from others. For compare the following. What makes someone a member of the human species? One obvious answer is: being born of human parents. This looks equally vulnerable to the charge of merely putting a problem off, on the grounds that there must have been at least two first humans

---

[6] Another reason is that most empiricists have regarded language use as somehow derivative, not being directly implicated in the processes of thought. For a brief critique of this view, see my *Tractarian Semantics*, ch. 10.

who were *not* human by virtue of having human parents. But in fact, as we now know, creatures that were recognizably human evolved gradually, in small steps, from creatures that were not. So it is possible that something similar may hold in the case of concept acquisition as well. It may be that what was recognizably a use of concepts evolved gradually, from the use of grunts and growls that were plainly non-conceptual. In which case an empiricist could explain concept acquisition in terms of linguistic training, without having to be committed to some form of abstractionism in explaining how the first concepts were acquired. But this is, so far, merely a promise. It remains for an empiricist to show how concepts *could* arise gradually out of something non-conceptual. We shall return to the issue in chapter 7. For the moment, it is enough to have noted the weaknesses in the classical empiricist accounts of concept acquisition.

Why be Anti-Nativist?

While the empiricist case against platonism is powerful, as we saw in chapter 3, its case against nativism is very weak by comparison. Not only are the direct arguments against nativism unsound, but the attempt to explain how all concepts may arise out of experience itself faces severe difficulties. This is not to say, of course, that nativism is then shown to be true. It is simply that the case against it is unproven. We may then remain puzzled as to why empiricists such as Locke and Hume should have been so convinced, nevertheless, that nativism must be false.

It might be replied that there is no special problem about this: they were simply misled by bad arguments. But I find this response unsatisfying. After all, Locke and Hume were both extremely intelligent men. So it remains possible, at least, that they may have had other, more powerful, reasons for rejecting nativism. Perhaps these may have gone unmentioned as a result of some sort of political expediency. For example, it may have been that their real reasons would have placed them in direct opposition to the Church.

Thus, one hypothesis might be that the early empiricists' rejection of nativism was part of their more general Enlightenment belief in the perfectability of man, and should be seen in contrast with the traditional Christian doctrine of original sin. But this proposal is hardly very satisfactory either, since there is no intrinsic connection between perfectability and anti-nativism. Enlightenment thinkers could equally well have maintained that while we have innately given knowledge, and innate faculties that are structured in such a way as to embody information about the world, our knowledge and attitudes nevertheless admit of indefinite extension and improvement.

It is true that if the mind were literally a 'blank slate', as the early empiricists seemed to maintain, then human nature would be almost unlimitedly malleable, for good or ill. The only constraints would be those of capacity (there may be limits to how much knowledge a human mind can contain, for example), and those imposed by the properties of the mental medium itself (some sorts of knowledge might be more difficult to acquire on the basis of general learning principles, for example). So the denial of nativism, if correct, would provide some sort of guarantee of human perfectability.

Endorsements of nativism, in contrast, would admit of at least two possible versions, implying either perfectability on the one hand, or inherent imperfection on the other. But even given a prior commitment to perfectability, this would be a very poor reason for rejecting nativism altogether. For it is just as plausible that the explanation of perfectability might be an innately given but indefinitely improvable nature. Moreover, the current proposal would require us to attribute to empiricists a belief in human perfectability that is apparently lacking in independent support, but that stands in need of it. For they cannot simply take for granted the falsity of Christian versions of nativism, unless they have some independent reason for rejecting nativism as such.

A rather different sort of proposal would be that the early empiricists' reasons for rejecting nativism might have formed such a fundamental part of their outlook as never to have been consciously articulated. Certainly it is common enough in philosophy for thinkers to allow themselves to overlook the weaknesses in their explicit arguments for a thesis, precisely because they have already become convinced of the truth of that thesis on other, and less readily articulable, grounds. Charity requires us to hope that something of this sort may be true in connection with the classical empiricist rejection of nativism. I shall return to the issue in the next chapter, and again in chapter 9.

5

Is Innate Knowledge even Possible?

In this chapter I shall consider various proposed accounts of the concept of knowledge, and how they bear on the question whether innate knowledge is possible.

A Justificationalist Argument

In chapter 4 we looked at Locke's reasons for denying that there is in fact any innate knowledge, and found them wanting. Some people have thought that he also had available to him a knock-down proof of the *impossibility* of innate knowledge, deriving from the concept of knowledge itself. But in fact this argument depends upon an account of the concept that is probably incorrect. Moreover, that Locke did not employ such an argument is evidence that he conceived of knowledge some-what differently, in a way that would allow innate knowledge to be possible in principle (even if it is not actual). I shall first explain the conception of knowledge from which the argument derives, and then develop the argument itself. I shall argue later that the account of knowledge in question is incorrect.

Many have held that knowledge can be defined as justified true belief. That is to say: in order to count as knowing something, you must at least *believe* it, what you believe must in fact be *true*, and you must have sufficiently good *reasons* for your belief. Such an account is not without its attractions. It is, for example, capable of explaining why someone who 'plays a hunch' that a particular horse will win a race does not know that it will win, even though the horse does in fact win just as they believed that it would. For such a person has insufficient reason for their belief; rather, it was a mere guess. The account can also explain why someone who has excellent reasons for their belief still does not have knowledge, if it should turn out from another, or later, perspective that their belief is not in fact true. For example, even someone who has just doped all the other horses in the race does not have knowledge that the remaining candidate will win, if it should turn out, through some bizarre sequence of accidents, that one of the others does.

(We can leave open, here, the question of just how powerful the reasons for believing some truth would have to be, in order to count as 'sufficiently good' to qualify that belief as knowledge. For essentially the same issue—of the degree of grounding required for knowledge—arises in connection with each of the various candidate accounts of the concept that we shall be considering. In fact I am inclined to think that the standards for a belief to count as knowledge are purpose-relative, varying depending upon what is at issue in the context in question. What counts as knowledge in response to a casual enquiry may not count as knowledge in a court of law, where someone's life may hang in the balance. So I suggest that 'sufficiently good' might mean 'good enough for the purposes in hand'. I shall return to this idea in chapter 11.)

If knowledge is justified true belief, then an innate belief can count as innate knowledge only if it has a justification that is also innate. This is because, in order for someone to know something on this account, it is not enough that there should *be* a justification. Rather, the person in question must possess that justification. For consider the following example. Playing a hunch, I decide that the horse in the red and green colours will win the race. But, unknown to me, there is in fact a sound justification for my choice—that horse has won all of her last ten races, whereas the rest of the field have just been doped. Even supposing that my horse does in fact win, I surely did not know that she would. For I was just guessing, and that is no basis for knowledge on any account of the matter. So even if there are innate beliefs, they will not count as innate knowledge if the evidence supporting them only emerges from subsequent experience.

Your reasons for a belief will presumably consist, in general, of other beliefs. If you have a reason for believing that world deforestation will be disastrous, then this must consist in some other belief or beliefs that imply or otherwise support that belief. And these beliefs, in their turn, must be justified if the original belief is to be justified. For you cannot provide justification for a belief through beliefs that are themselves *un*justified. So it looks as if we have a regress: in order for one innate belief to be innately known, there must be some other innate belief that justifies it; and this in turn must have a justification that is also innate, and so on to infinity. In which case innate knowledge is impossible. For we surely cannot have infinitely many innate beliefs.

It might be said that this regress could terminate with beliefs that are *self-evident*, that provide their own justification. This may be so. But in fact there are only two sorts of belief that are plausible candidates for such self-evidence, and neither of these kinds will be innate. First, a belief may be self-evident in virtue of concerning the subject's own immediate experiences (such as the belief that I now feel a pain). But in this sort of case it would be absurd to suppose that the belief in question is innate. Secondly, simple truths of reason may be self-evident, such as the belief that $2 + 2 = 4$. But in fact such beliefs will only count as self-justified if they concern nothing beyond themselves (that is, if they are *analytic*). For, in this context, self-evidence cannot be merely a matter of a belief striking us as intuitively obvious, since there would remain the question of what reason there is for thinking that such obviousness is a good guide to the state of the world in question.

To see this, suppose that the subject-matter of the proposition '$2 + 2 = 4$' is construed platonistically, as concerning necessarily existing abstract individuals. Although this proposition may seem to us undeniable, this is in fact insufficient for it to count as self-justifying. For what reason have we for supposing that obviousness-to-us (a feature of our psychology) is a good guide to the way things really are in the abstract realm? If we lack any reason for this belief, then the proposition will not really be justified, despite our inability to deny it. And if we possess such a reason, then the justification for the proposition will derive from some other belief of ours, and the regress will continue. So innate self-evident propositions can only halt the regress if their subject-matter concerns nothing beyond themselves (that is, if they are analytic).

What this argument appears to show is that it is impossible that there should be innate knowledge that is at the same time substantive (that concerns a reality that is independent of our minds). This is not quite the same as saying that innate knowledge is impossible altogether. But in fact, given any developmental construal of innateness (two versions of which were outlined in chapter 4), that a belief is analytic will undercut all warrant for supposing it to be innate. Only if we were prepared to believe that analytic propositions are stored in the mind as such from birth could we accept that there is innate knowledge of any sort, if the argument above is sound.

The reason why acquired analytic beliefs would not be counted as innate is that there will be no problem in explaining how people might come to possess them through general learning mechanisms. Since knowledge of simple (self-evident) analytic truths requires only that we be capable of discerning simple relations between our own concepts (ideas), there will be insufficient reason for saying that the knowledge in question is innate. Even if the constituent concepts are innate, our knowledge of the relations between them need not be. For in order to explain how we come to possess such knowledge, we only need to suppose that we have the general ability to compare and contrast our own ideas.

A Coherentist Suggestion

It should be noted that the argument developed above against the possibility of innate knowledge presupposes the truth of foundationalism in the theory of knowledge. For it assumes that the justification-relation between beliefs is a *linear* one. The argument takes for granted that one belief will be justified by another, which is justified by another, and so on until we reach beliefs that somehow justify themselves (either reports of immediate experience or simple analytic truths), these latter beliefs constituting the foundation for the rest. But what if we were to embrace a coherentist account of justification instead? What if we were to take the view that a belief may be justified by virtue of forming part of a mutually supporting network of beliefs, each member of which receives its justification from its relationship with the others? This would be to picture know-ledge, not as a pyramid built on secure foundations, but rather as a web held together by the relationships between its component parts. If this account were adequate, then there would be no reason why we could not possess innate know-ledge. For we might be born with (or born determined to develop) a set of consistent and mutually supporting beliefs on some subject-matter. In which case, provided that those beliefs were true, we could be said to have innate knowledge. For given a coherentist conception of justification, those innate beliefs would not only be true but justified.

Coherentism faces considerable difficulties if taken at face value, however. For there may be any number of coherent bodies of beliefs that are mutually inconsistent with one another. For example, suppose that both Christianity and Hinduism are internally coherent. Yet they cannot both be true, since there cannot be both one God and many. But in that case it is very hard indeed to see how the mere fact of the coherence of Christian belief could render it justified, given that there are other incompatible beliefs that are equally coherent. For there would then be no justification for being a Christian *rather than* a Hindu.

Even if coherence is construed in such a way that a body of beliefs must be more coherent than any incompatible set of beliefs in order to count as justified, there will still be problems. For suppose that I had read Jane Austen's *Emma* as a child, under the mistaken impression that it was the biography of a historical individual. Then as an adult I still retain a great many beliefs about Emma Woodhouse, but have forgotten how I came to have them. Since these beliefs constitute a consistent, mutually supporting set, a coherentist seems committed to saying that they are justified. But this is counter-intuitive. I think we should be strongly inclined to deny that a coherent set of fictional beliefs could count as justified in this example.

Coherentists are no doubt correct to stress, as against some (but not all) forms of foundationalism, that a justification for a belief can appeal to non-self-evident principles, such as infer-ence to the best overall explanation of a given range of phenomena. But the beliefs within an explanatory network must surely be anchored somewhere. They cannot just 'float free' of all constraint, as do my beliefs about Emma Woodhouse in the example above. On the contrary, I suggest that there must be some beliefs given to us in experience, providing the data that the coherent network explains. Just such a weakened form of coherentism (which may alternatively be seen as a weakened form of foundationalism) will be considered in more detail in chapter 12.

If some suitably weakened form of coherentism is correct, then it will certainly be possible that there should be innate knowledge. For suppose that some set of true beliefs were innate, serving to provide a coherent explanation of the data provided by subsequent experience. Then provided that those beliefs could not have been (or at least were not) *learned from* the experiences in question, they would count as innately known. They would be true beliefs whose justification is triggered by experience (either generally or locally), rather than derived from it.

This suggestion cannot, however, serve to explain why Locke (and other empiricists) failed to deploy the argument against the possibility of innate knowledge sketched earlier. For it seems certain that none of the classical protagonists in this debate— whether empiricist or rationalist—endorsed a coherentist conception of justification. A better explanation is that Locke employed some other account of knowledge, which would leave open the possibility of innate knowledge even when conjoined with foundationalism. This idea is also supported by considerations of charity, since there are powerful reasons for thinking that the justificationalist conception of knowledge is incorrect. To this issue I now turn.

Against Justificationalism

While a conception of knowledge as justified true belief may raise a difficulty for the possibility of innate knowledge, that conception has itself come under increasing pressure in recent decades. One problem with it is that we are then constrained to deny knowledge to those, such as children, who may be unaware of the justification for their true beliefs. Yet it is counter-intuitive to insist that a child does not know that Father is washing up dishes (even though the child can see that he is doing so), on the grounds that the child is as yet incapable of providing any sort of justification for its belief. (After all, the child may be just as reliable in reporting such matters as the rest of us.) Equally, suppose that you are unable to respond adequately to a sceptic who demands to know what reason you have for believing that you are not now dreaming. Then if justificationalism is correct, you do not in fact know that you are awake. It will only be the privileged few, who can answer the sceptic, who may be said to know that they are not dreaming; the rest of humankind, in lacking a justification, will lack knowledge also.[1]

This problem is really very widespread. For few of us are capable of articulating anything like a convincing justification for more than a handful of our beliefs. (Try asking an ordinary person for their reasons for one of their everyday beliefs—for example, their belief as to the date of their mother's birthday—and see how quickly their feeble attempts at justification will run out.) Yet we cannot avoid the difficulty by saying that knowledge only requires that there *be* a justification, not that the subject be aware of it. For, as we saw earlier, this would then permit many beliefs that are really just guesses to count as knowledge. Neither will it help to say that knowledge only requires that a justification be *constructable* from amongst the subject's other beliefs, not that the subject need actually, themselves, have effected such a construction. For suppose that I had laid a bet on the winning horse because I liked the colours in which her jockey was dressed. This would surely be a mere guess rather than knowledge, even if I did in fact know (but failed to recall) that she had won all of her last ten races.

Not only does justificationalism conflict with our common-sense views about the circumstances in which people may be said to possess knowledge, but it is far from clear why we should *want* to insist that knowledge requires justification. For our main interest in the question whether or not someone knows something is that a positive answer will warrant us in adopting that belief for ourselves. And while the fact that they are justified in holding their belief may be sufficient to give us such a warrant, it is unclear why it should be necessary.

---

[1] I here assume that there *is* an answer to the sceptic. See chapters 11 and 12.

The main use we have for the concept of knowledge is that if someone knows something, then we may safely add what they know to our own stock of beliefs. We therefore need to be able to establish that they have knowledge independently of assessing, for ourselves, the truth of their belief. (If we could do that, we should not *need* to know that they know.) And the basic fact we need to establish is simply that their belief is, in the circumstances, *likely* to be *true*. Whether or not they themselves can provide reasons for their belief is of no particular importance. All that really matters is that the believer should, in the circumstances, be *reliable* (or reliable enough) on the topic in hand. Indeed, it seems to me that from the point of view of our practical interest in knowledge, the appropriate concept is a reliabilist one, of the sort to be defended in the next section.[2]

Even if justified true belief is not necessary for knowledge, it may still be sufficient. But this thesis, too, is vulnerable to counter-example. Thus suppose that I had turned on my radio in order to hear a live account of the 1988 Olympic 100-metre final, having just read in the paper that such a programme was due to be broadcast at that time. I duly hear the commentator describe a race in which Ben Johnson beats Carl Lewis, setting a world record in the process—which is exactly what was in fact happening. But unknown to me there was a fault in the satellite transmission from Seoul, and they were in fact broadcasting a repeat commentary from the previous year's World Championships instead, in which a similar race took place. Here I have a justified true belief that Johnson has just set a world record in beating Lewis, but I surely do not know it. There are many examples of this general sort.[3]

What this example shows is at least that the account of knowledge as justified true belief needs to be supplemented in some way. Perhaps we might require in addition that the belief should be non-accidentally connected with the fact that it concerns. For the distinctive feature of the example is that the

---

[2] While I otherwise concur with Edward Craig in 'The Practical Explication of Knowledge', *Proceedings of the Aristotelian Society*, 1987, he seems to me to go wrong on just this point. For a defence of the general approach to conceptual issues exemplified both here and in Craig's paper, see my 'Conceptual Pragmatism', *Synthese*, 1987.

[3] They are generally known as 'Gettier examples' after E. Gettier author of 'Is Justified True Belief Knowledge?', *Analysis*, 1963.

truth of the belief in question is entirely fortuitous, despite being justified. But our earlier arguments (concerning the knowledge of children, and your knowledge that you are now awake) suggest something much more radical. For if a belief can count as knowledge although the believer lacks any justification for it, then we should reject the justification-condition altogether, replacing it with some other condition.

Causal Theories and Reliabilism

Some have developed an account of knowledge that is causal in character. They have said that knowledge is true belief that is caused by the fact which it concerns.[4] This can explain why, in the example above, my belief that Johnson has broken the world record does not count as a case of knowledge. For the fact of his having broken the record fails to play any role in the causation of my belief. The account can also explain why the child knows that its father is washing up, despite lacking a justification. For the child's belief is in fact caused by perceiving Father at work at the sink. Similarly, you may be said to know that you are not now dreaming, despite your inability to answer the sceptic, provided that your belief that you are not dreaming is in fact caused by the fact that you are awake.

Although the causal theory of knowledge is in many ways attractive, it runs into trouble if it is to allow for the possibility of knowledge of the future, and of knowledge of unrestrictedly general statements. For future states of affairs surely cannot be causes of present belief, and it is implausible to suggest that our beliefs about general laws of nature are in fact caused by those laws themselves. Thus suppose that I have just set light to the fuse on a firework. I know that it is a rocket of reliable manufacture, which has never failed in the past. I know that it has been stored in dry conditions, that the weather itself is now dry, and that there is no wind. Surely in these conditions I may know that the rocket will shortly take off. But it is not the fact that the rocket will take off that causes my belief that it will (as

[4] This is roughly the position held by Robert Nozick, *Philosophical Explanations* (Oxford: Oxford University Press, 1981), ch. 3.

the causal theory of knowledge would require). This would involve backwards causation, which is impossible. My belief is rather caused by the facts that I have mentioned, such as my belief that the firework is dry. Equally, consider my belief that all massive bodies attract one another (the law of gravity). This may surely count as knowledge. But my belief is not caused by the fact that *all* bodies attract (past, future, and distant), but rather by the bodies that I have observed, and by the reports of other observers and scientists. So again, we appear to have a case of knowledge without causation by the fact which it concerns.

We can keep all the advantages of the causal theory, while avoiding its difficulties, if we say that knowledge is true belief that is *caused by a reliable process*. A reliable process is one that generally issues in true beliefs, and that also serves, in the particular case in hand, to discriminate reliably truth from relevant falsehood.[5]

This can explain how we may have knowledge of the future, if the process of inference from past facts and present tendencies is (in the circumstances) a reliable one. It can also account for our knowledge of laws of nature, if the processes of induction and inference to the best explanation employed by scientists are also generally reliable. Yet we can still explain why I would lack knowledge of Johnson's victory, in the example discussed above. For while listening to radio broadcasts may be a generally reliable method of acquiring true beliefs, in this case my belief would have been the same even if Johnson had *not* won. In the circumstances the process that causes my belief is not reliable in discriminating truth from relevant falsehood. We can also still explain the fact that the child knows its father is washing up, as well as your knowledge that you are not now dreaming. For the child's perceptions in circumstances such as these (with the observed events taking place not too far away, in good lighting, and not too swiftly) will generally result in true beliefs. And your belief that you are now awake results from a process that involves your current conscious awareness, which is also reliable.

[5] For a detailed exposition and defence of reliabilism, see Alvin Goldman, *Epistemology and Cognition* (Cambridge, Mass.: Harvard University Press, 1986), ch. 3.

While the thesis that knowledge implies causation by a reliable process is beginning to look plausible, it might be claimed that we have been too hasty in dropping the justification-condition. For consider the following example.[6] I am taking part in a psychological experiment, where the experimenter has told me that I have been given a drug that will totally distort my visual perception. Nevertheless, I perversely continue to believe, on the basis of my perceptions, that there is a pink rabbit sitting on the desk in front of me. But in fact the experimenter has given me a placebo, and there really is a pink rabbit on the desk, which has escaped from the genetics laboratory next door. This is a case in which my belief appears to be caused by a reliable process (undistorted perception), and is in fact true. But, it may be urged, I surely do not *know* that there is a rabbit on the desk. This is because it is, in the circumstances, unreasonable of me to trust my visual perceptions.

In fact, however, the correct response to this example is that I *do* have knowledge that there is a pink rabbit on the desk before me. This is because I am, in the circumstances, a reliable informant on such matters. The intuition that I do not have knowledge only arises through a confusion of levels. It is true that, representing the example to myself in the first person singular, *I* cannot correctly claim to have knowledge of the presence of the rabbit. But others may correctly claim this of me, and I may correctly claim it of myself at a later time when the details of the situation emerge. This will be made clear in the next section.

Orders of Knowledge

Reliabilism can only be acceptable if we are prepared to deny what is sometimes called 'the KK thesis', which holds that in order to know something, you must at the same time know that you know it. For the process that gives rise to a belief can in fact

---

[6] The example is a development of Goldman's in *Epistemology and Cognition*, 53. He uses it to argue for the inclusion of a justification-condition in the account of knowledge, which he then tries to cash out in a reliabilist way. I shall argue in the next section that he is wrong on both counts. I once deployed a similar example myself in defence of justificationalism. See my *Introducing Persons* (London: Routledge, 1986), 6. i. A.

be reliable even though you yourself do not know that it is; indeed, you may not even know what that process *is*. Then according to reliabilism your belief will count as knowledge, but you will not know that it is knowledge.

There is some reason to think that it is the KK thesis that gives rise to justificationalism. For if knowing something requires that you know that you know it, then you will only have knowledge if you know that your beliefs have been formed in a way that renders them likely to be true— in other words, if you possess a justification for the belief. But why should the KK thesis be accepted? Surely knowing (first-order knowledge) is one thing, knowing that you know (second-order knowledge) is another. Indeed, the thesis threatens to degenerate into a regress. For if taken generally, then you will only know that you know something, in turn, if you know that you know that you know it; and so on.

It seems to me likely that both justificationalism and the KK thesis result from conflating our basic interest as epistemologists (theorists of knowledge) with what it is that we take an interest *in*, namely knowledge. I propose to argue for this in a number of steps. First, I claim that the central question of epistemology is what we may know ourselves to know (a second-order question). This is easily overlooked. For if I ask myself 'What do I know?', the term 'know' only figures in the question once. But in fact, in returning a positive answer to the question whether I know that such-and-such, I should be tacitly claiming to know that I know it. For suppose that I assert 'The ozone layer is shrinking'. This is in fact a claim to *know* that the ozone layer is shrinking. Any unconditional assertion is, in effect, a tacit knowledge-claim. For to respond to it by saying 'But you don't really *know* that' is to challenge the speaker's right to say what they have said. Similarly, then, if I assert 'I know that the ozone layer is shrinking'. This is in fact a tacit claim to know that I know it.

My second thesis is that the only generally reliable method of acquiring second-order beliefs about what I know (that is, of acquiring second-order knowledge) is by a process of reasoning. The only way to know that you know, is to construct a justification for claiming to have knowledge. There could not, for example, be any process of introspection that would be reliable in obtaining for us beliefs about our own states of knowledge. For while introspection might have access to my first-order states of *belief*, it cannot have access to the fact that those beliefs were produced by a reliable process, when they were. It follows that when we come to do epistemology (seeking to know what we know), justificationalism and reliabilism converge: each will arrive, by different routes, at the view that we need to seek justification for our claims to knowledge.[7] It would then be quite natural that philosophers should mistakenly slip into thinking that knowledge itself (whether first-order or second-order) requires justification.

This, then, is how we should respond to the example of the pink rabbit presented earlier: since my belief in the presence of a rabbit is in fact produced by a reliable process, I may truly be said (by others or by myself at a later time) to have knowledge. For someone who knows that I have been given a placebo (and so who knows that my perception will most likely be accurate) may reasonably treat me as a reliable informant on the matter. It is only when I raise the question about myself at the time—'Do I know that there is a pink rabbit here?'—that the answer has to be negative. For the question is, tacitly, a second-order one. An affirmative answer will claim to know that I know. But, as we have just seen, such second-order claims have to be based on a process of reasoning. And it is plain that, in the circumstances, I lack any justification for saying that I know.

It is important to note that the relevant concept of justification, in this context, has to do with the reasons that are available to the person whose knowledge is in question. Whether or not a belief is justified has to be determinable from the perspective of the subjects themselves.[8] This is because the main use we have for the concept of justification is in determining what we ourselves should believe. The question whether someone else is justified in holding a belief is generally of very little interest, except in so far as believing on the basis of a justification is one form of reliable belief acquisition, or except by way of providing a moral applicable to our own case. We therefore want a concept whose conditions of application are defined in such a way as to

---

[7] This point will prove to be of some importance in chapters 11 and 12, when we come to consider the problem of scepticism.

[8] This is where Goldman's version of reliabilism seems to me to go badly wrong, in making justification a matter of objective reliability.

be available to us. (This is in contrast with the case of knowledge itself, where our main use for the concept is a third-person one —on the question as to whether someone knows something will turn whether or not I myself should believe that thing. Here all that matters is that the person in question should be reliable on the issue at hand.)

In the light of the above considerations it seems to me quite possible that classical philosophers such as Locke and Descartes may have been implicitly working with a reliabilist conception of knowledge, despite their overt concern with justification. For as epistemologists they were interested in the question of what we could know ourselves to know. To seek an answer to this question is to look for reasons in support of our various (first-order) claims to knowledge. But the fact that the process of acquiring second-order knowledge of this sort involves justification is quite consistent with the thesis that knowledge itself is reliably acquired true belief.

Indeed, Descartes's strategy in the *Meditations*, for example, is quite naturally interpreted in this light. For what he wishes to know is whether the processes through which he acquired his beliefs were reliable. He believes that he can answer this question affirmatively, by virtue of proving the existence of a veracious God. But there is no suggestion that ordinary people do not have perceptual knowledge until they, too, have grasped such a proof (that is, until they, too, have provided a justification for their knowledge-claims). Rather, they have knowledge because perception is *in fact* reliable. What they lack is only the knowledge that they know.

Innateness and the A Priori

If reliabilism is acceptable as an account of knowledge, then the possibility of innate knowledge is left open. Innate beliefs will count as known provided that the process through which they come to be innate is a reliable one (provided, that is, that the process tends to generate beliefs that are true). There are two possible candidates for such a process: divine intervention on the one hand, and evolution on the other. We could maintain, as most classical rationalists did, that innate beliefs are directly implanted in the mind by a veracious God. Or we could hold that innate beliefs have been acquired through evolution, via natural selection. Each of these candidate processes would most probably be reliable. We shall return to consider them in more detail in later chapters.

I have argued for a conception of knowledge that would allow innate knowledge to be possible, no matter whether we have a foundationalist or a coherentist conception of justification. And I have suggested that the reason why Locke did not employ the argument against innate knowledge sketched at the outset of this chapter is that he too may have conceived of knowledge in reliabilist terms. But would innate knowledge, if it existed, at the same time be a priori? In one sense at least it would be. For, as we noted in chapter 1, one thing that can be meant by saying that something is known a priori is that our knowledge of it is independent of empirical support. In this sense, a priori knowledge is knowledge that has not been learned from experience. And innate knowledge (reliably caused innate belief) would apparently possess just such a status, neither having been learned from experience, nor requiring support from experience to qualify as knowledge.

However, to say that innate knowledge would be a priori is one thing, to say that we may know that we know it a priori is quite another. Knowledge that is a priori in this sense will not, unlike what is generally held concerning the other main traditional category of the a priori, be knowable to be such by a process of thought alone. On the contrary, to know that a belief is both innate and reliably produced, and hence that a given item of knowledge is a priori, may require argument from empirical premisses. An appropriate defence of nativism will not itself be a priori but broadly empirical. This topic—the defence of nativism—will now occupy us over the next three chapters.

6

The Case for Innate Mental Structure

Having considered the empiricist case against nativism and found it wanting, I shall now begin to assess the evidence that supports nativism.

Principles of Learning

All empiricists have allowed that the mind is innately structured into distinct faculties, including for example thought, perception, and memory. And all have allowed that the basic mechanisms involved in the acquisition of our beliefs—such as the recording of experiences in memory, and the laws of association amongst ideas—are innately given. (One cannot, from nothing, learn how to learn.) But they have denied that the mind divides into faculties with innate constituent structures that are distinct from one another, embodying information about the domains they concern. On the contrary, they have insisted that all belief-acquisition principles are *general* ones, operating in similar ways for all domains of knowledge. Much of the evidence that has emerged over recent decades, on the other hand, suggests that there is innate information embodied in the language faculty, for example, that is not present in the faculty of vision or vice versa. (I shall hold over until later, discussion of the question whether such information may be described appropriately as 'knowledge'.)

In fact there might appear to be two different issues here. First, there is the question whether the various mental faculties employ distinct innate belief-acquisition principles. Secondly, there is the question whether our innate mechanisms of belief acquisition can be said to embody information about the world.

It is easy enough to see why empiricists should have denied the latter possibility. For even if such information could not itself be counted as genuine knowledge, it would plainly contribute to our knowledge, and might thus be expected to fall within the scope of the empiricist rejection of nativism. However, it is not nearly so obvious why empiricists should have insisted that all belief-acquisition principles are general ones, the same for all domains of knowledge.

One possible explanation lies in the conception of science endorsed by the early empiricists.[1] For the assumption within the physical sciences at the time was that all physical phenomena should be explicable in terms of laws of nature that are entirely general, applying equally to all interactions and combinations of matter. In effect, it was assumed that all laws of nature that concern the physical realm should be reducible to the laws of physics. Now, as well as being epistemologists, the early empiricists certainly saw themselves as attempting to develop a science of the mind (more on this in chapter 9). So quite probably they would have imported the above assumption into the domain of psychology, thus insisting that all mental processes should be explicable in terms of the same general psychological mechanisms, such as laws of association amongst ideas.

Few thinkers today would endorse the reductionism inherent in early conceptions of science. Few would now insist that all sciences that deal with physical phenomena, including chemistry, biology, and physiology (and since most scientists today are materialists, they would also add psychology), must in principle be reducible to physics. On the contrary, most would allow that the special sciences can contain laws that, while they may (indeed must) be *consistent* with the laws of physics, cannot be *deduced from* them.[2] Then if the laws of the special sciences are autonomous, there is no reason to insist, either, that within any given science, such as psychology, all phenomena must be explicable in terms of a single set of laws. We therefore seem to have no reason, today, for insisting that the psychological

---

[1] On this see A. D. Smith, 'On Primary and Secondary Qualities', *Philosophical Review*, 1990.

[2] See e.g. Jerry Fodor, *Representations* (Brighton: Harvester, 1981), ch. 5.

mechanisms operative in the various faculties of the mind should be the same throughout.

But in fact there is also a more direct explanation for the empiricist insistence that learning mechanisms should be the same for all domains of knowledge. For if those mechanisms were to differ in different cases, this could only mean that they embodied information about the domains that they concerned. Suppose, for example, that the mechanisms through which we learn the grammar of our native language were to differ from the mechanisms though which we acquire our common-sense beliefs about material objects. It is then hard to see how this could be so—to see why one could not, for example, employ the one set of learning principles to acquire knowledge of the other domain—unless the mechanisms in question contained, at least tacitly, information specific to the domains that they concerned. In which case the insistence on general learning principles may be seen to derive from the empiricist view that mechanisms of learning should not in any way embody information about the world.

We now need to consider whether empiricism is correct in this regard. I shall focus first on the case for an innately structured language faculty, considering in the final section the case for an innately structured visual faculty.

Chomsky on Language

Noam Chomsky has been prominent in arguing that human beings possess a distinct language faculty, which is involved in the acquisition and use of natural language. The principles of operation of this faculty are said to be distinct from those of other psychological faculties, containing a good deal of information about natural languages.[3] In fact Chomsky's view is that much of our linguistic knowledge is innate, being embodied in the structure of the language faculty. When a child acquires its first language, this is not so much a matter of learning as of the

[3] See e.g. his *Language and Problems of Knowledge* (Cambridge, Mass.: MIT Press, 1988). For a more detailed, but still introductory, exposition of Chomsky's ideas than that given here, see V. J. Cook, *Chomsky's Universal Grammar* (Oxford: Blackwell, 1988).

language faculty being triggered into spontaneous growth. Chomsky concedes, of course, that some exposure to English is necessary if the child is to grow up speaking that language rather than French or Japanese. But he denies that this is all a matter of learning—the child only has to learn the words of the language (the lexicon), not the grammar. He is therefore committed to the theory of local triggering of innate knowledge, which we distinguished from other varieties of nativism in chapter 4.

It might be objected that there can be no way of testing Chomsky's hypothesis. For once we allow that innate knowledge may be locally triggered by experience of the domain that it concerns, how are we to tell, in any particular instance, whether the knowledge has been learned or triggered? But, in fact, the general point about a mechanism of learning is that the final state of the system should be a direct product of the initial experience, co-varying with it. In contrast, a mechanism of local triggering will malfunction in the face of initial experiences for which its operative assumptions fail to hold. So if our knowledge of the grammatical basis of natural languages is locally triggered, then we may predict that a child exposed only to an artificially constructed language will either fail to learn it at all, or will only do so extremely slowly by comparison with normal language acquisition. However, as this example makes clear, direct testing of the issue may be ruled out on moral grounds. In settling the question whether some item of knowledge is learned or locally triggered, we shall often have to rely upon indirect arguments.

However, this need not mean that we cannot sometimes form reasonable beliefs on the matter. In particular, wherever it is hard to see how our knowledge of some domain could have been acquired from our experience via any combination of memory, inductive extrapolation, and inference to the best explanation, then we shall have reason to believe that local triggering has occurred. As we shall see in the next section, this is just what Chomsky claims. He insists that a child's mature knowledge of its native language will go well beyond anything that it could have learned from its limited exposure to that language.

Crucial in the development of Chomsky's ideas has been the

discovery of what he calls 'linguistic universals'. These are abstract features of syntax that have been found to be common to all natural languages.[4] For example, all known languages have a syntax that is dependent upon phrase structure rather than, for example, linear word order. And the basic form of sentence structure within all natural languages is subject-phrase–verb-phrase (where the object of the verb is incorporated into the verb-phrase), rather than the form subject-phrase–verb-phrase–object-phrase employed in the artificial languages of logic. Chomsky's view is that these universals reflect our common genetic endowment, given in the structure of the language faculty.

Moreover, even where there are manifest syntactical differences between languages, Chomsky thinks that these can be seen as merely different settings in a single underlying innate parameter.[5] In his view, acquiring such a feature of syntax will be a matter, not of learning, but of switching the parameter to the appropriate setting. Of course Chomsky concedes that the lexicon (the individual words) of one's language need to be learned, together with some specific rules of grammar. But he maintains that this is largely a matter of learning the labels for a set of innate concepts, most concepts themselves being *un*learned. I propose to set aside his views on concepts to the next chapter, concentrating here on his nativism in relation to syntax.

One argument for nativism concerns the explanation of the existence of linguistic universals themselves. For why should all natural languages have features in common, unless this reflects the innate structure of a language faculty that we all possess? Certainly there does not appear to be any general feature of our psychology, nor any general learning principle, that could explain why the syntax of all languages is phrase structure-dependent. However, an equally plausible explanation, alternative to nativism, exists in the hypothesis of a common origin. Let us suppose, as is likely, that all humankind is descended from a common stock. Then there will probably have been a

[4] To believe in linguistic universals need not, of course, mean believing that they have independent existence—it need not mean accepting any form of platonism.

[5] These parameters may usefully be thought of as disjunctive universals, taking the form 'Natural languages are constructed *either* like this, *or* like this, *or* like this'.

time in the distant past when all living human beings spoke the same language. As different groups then dispersed around the globe, their language would have begun to change and develop, in perhaps radically different directions; but always, it may be supposed, retaining certain general features in common— today's linguistic universals.

Plato's Problem

In fact Chomsky himself places greater emphasis on the argument from what he calls 'Plato's Problem' (so named after the slave-boy example in Plato's *Meno*). The problem is essentially this: how, in the absence of innateness, would children manage to learn so *much*, so *fast*, and on the basis of such *meagre exposure* to the language? The issue of quantity learned is of course impossible to measure precisely, but some sense of it can be gained by glancing at any contemporary work in linguistics. Natural languages seem to be governed by a great many different syntactical rules of very considerable complexity. While these rules are not consciously known by a native speaker of the language, they seem nevertheless to implicitly govern linguistic behaviour. They emerge, for example, in the speaker's judgements concerning what constructions are or are not permissible in particular cases. The question then is: how does the child manage to discover all these rules within the space of just a few years, employing only learning mechanisms, when even many years of co-operative labour on the part of linguists still leaves much of the grammar of natural languages controversial?

The problem becomes even more pressing when one considers the nature of the data available to the child. Adult linguists, who may themselves be native speakers, will have available to them as data in the construction of their grammars a potential infinity of sentences of the language. They can call up sentences at will, and judge for themselves whether they are permissible on the one hand, or ill formed on the other. The child, in contrast, has to arrive at its knowledge of the grammar of its native language on the basis of the very limited sample of sentences to which it happens to have been exposed. Moreover, it will have, in general, only positive data to go on. No one explicitly tells the child that certain constructions are impermissible. Nor, when the child itself speaks, are its errors often corrected. In general people simply speak to children, without attempting to *tell* them how to speak; and they allow children to speak for themselves, only very rarely correcting their mistakes. Even more strikingly, indeed, much of the data available to the child is actually corrupt (that is, ungrammatical). To see this, try taping part of an ordinary conversation; transcribe it on to paper, and then study the result. What you will find is that much of the conversation will consist of sentences that are never finished, sentences that start out in one direction and then conclude in another, and so on. Yet on this sort of basis the child acquires a mastery of the grammatical rules of the language, which still eludes complete description by linguists.

This argument of Chomsky's appears extremely powerful. It is impossible to see how the child could construct a non-conscious model of the grammar of the language, using only general learning principles (such as analogy, inductive extrapolation, inference to the best explanation, and so on), on the basis of such fragmentary and corrupt data. Nor is it easy to see how there could be any learning mechanism specific to the language faculty that could generate knowledge of grammar from such data without presupposing anything about the structures of natural languages. It appears that Plato's Problem in the case of language can only be solved if we suppose that much of the child's grammatical knowledge is already innate. The child would then only have to learn the lexicon of its language and a few grammatical rules that are language-specific, provided that it has sufficient exposure to instances of sentences of the language to set the various parameters within its innate language faculty. This looks like a manageable task.

There is a particular development of the above argument that Chomsky also employs. This is that children simply do not make the sorts of mistake that one would expect them to, if they were arriving at their knowledge of the language through general learning principles. Consider, for example, how the child learns to construct questions from the corresponding declarative sentences. It first has experience of simple sentences such as 'The man is at home' and 'The man is happy', noticing that the corresponding questions are 'Is the man at home?' and 'Is the man happy?' respectively. At this stage the simplest, most obvious candidate for the rule of question formation is to remove the *first* occurrence of the word 'is' (or its cognates) to the front of the sentence, leaving the sentence otherwise unchanged. One might then expect that when faced with the task of constructing the question corresponding to the sentence 'The man, who is happy, is at home', a child might try 'Is the man, who happy, is at home?' But no child ever says such a thing! The next most obvious candidate for the rule would be to remove the *final* occurrence of the word 'is' to the front of the sentence, leaving the sentence otherwise unchanged. But applying this rule to the sentence 'The man is at home, who is happy' would produce 'Is the man is at home, who happy?' But no child ever makes this sort of mistake either!

Of course the correct rule, as Chomsky points out, is not linear, as these are, but phrase structure-dependent. The rule is that one should remove the occurrence of 'is' from the main verb (the verb of the main verb-phrase) to the front of the sentence. For in reality the sentence 'The man, who is happy, is at home' has the structure '[[The man] [who is happy]] is at home', where the outer brackets enclose the subject-phrase. Similarly, the sentence 'The man is at home, who is happy' can be represented as having the form '[The man] is at home [who is happy]', where the brackets collectively enclose the subject-phrase. Since these structures are amongst the last that one might have expected, had one been employing only general learning principles, the information that natural languages are phrase structure-dependent (as well as much else besides) must be supposed to be innate to the language faculty.

Someone might be inclined to object against Chomsky that grammar (the grammar of English, for example) *is* explictly taught, often in schools. Indeed, are not the letter columns of newspapers full of complaints that grammar is *inadequately* taught? How, then, can Chomsky claim that children acquire a near-perfect grasp of it by an early age? But this is a misunderstanding. In the view of Chomsky, as of most contemporary linguists, the various dialects of English constitute autonomous languages, with their own rules of sentence construction. It is only for political reasons that we select one of these dialects as primary, and try to impose it uniformly across all sections of the community. So Chomsky's reply to the objection would be that children do indeed have a near-perfect grasp of the grammar of their native English dialect at an early age; what they are then required to master in school is the grammar of *another* dialect, which they have to learn almost as they learn a foreign language.

Are Grammars Known?

We may agree that in some sense or other a mature speaker of a language has succeeded in internalizing the grammar of that language. For such a speaker has the ability to tell, of any candidate sentence of the language, whether or not it is syntactically well formed. There must then be an internal structure that explains this ability, in some way corresponding to the rules constructed by linguists. But is it appropriate to describe the ordinary speaker as *knowing* the grammar of their language? Should we say that the innate structure of the language faculty provides us with locally triggered innate *knowledge* of the syntactical rules governing our native language? Chomsky argues at some length that we should, insisting that we are here dealing with a genuine species of innate—although non-conscious—knowledge.[6]

It is worth noticing that if we accept the position argued for in the previous section, then in any case we shall be provided with *some* locally triggered innate knowledge. For example, your knowledge in the particular case that 'Is the man, who happy, is at home?' is not a permissible question-form would constitute just such knowledge. For you know this without being told, and without having learned it from your experience of the language. This is not at issue. What *is* at issue is whether the underlying systems of rules, which generate this particular item of knowledge, may be said to be known in their turn.

Chomsky contrasts his claim that we have knowledge of grammatical rules with the idea that our grasp of them constitutes, rather, a *practical capacity*, or skill. He presents a number of arguments against the latter hypothesis. In the first

[6] See *Language and Problems of Knowledge*, ch. 1.

place he points out that two people can have exactly the same knowledge of English, while differing greatly in their ability to use it. For example, one may be an accomplished and creative public speaker, whereas the speech of the other is pedestrian and banal. Yet they each have the same knowledge of the grammar of English, for example agreeing as to which sentences are or are not well formed. But here Chomsky has picked on a different ability to the one intended. The skill that Chomsky's opponent contrasts with our supposed knowledge of the grammar of English is not the ability to use English well, or creatively. It is simply the ability to produce, and to recognize, sentences of English that are grammatical. And in this respect there may be no difference whatever between the creative and the dull speaker.

Chomsky also cites the case of people who suffer complete aphasia after a head-wound, losing all ability to speak or to understand their native language. But then, as the effects of the injury recede, they may recover their ability to understand without any relevant instruction or experience. What this shows, Chomsky thinks, is that since people can retain their knowledge of the language while losing their ability to use it, our grasp of grammatical rules cannot simply be a matter of ability, but is a genuine instance of knowledge. What he fails to notice, however, is that one can retain the *categorical basis* of a capacity, while circumstances prevent that capacity from being exercised. (To say that something has a capacity is to say that it *will* behave in a certain way *if* certain other things happen. The categorical basis of a capacity is the positive—non-hypothetical —fact that explains *why* the thing would behave in that way in those hypothetical circumstances.)

For example, consider the brittleness of a glass, which is a matter of its capacity to be broken easily (it *will* break *if* you hit it). This capacity has a categorical basis, presumably in the molecular structure of the glass, which explains why even modestly severe impact will shatter it. Now suppose that the glass has been tightly packed in cotton wool. Of course it retains its brittleness (its molecular structure remains unchanged). But as a matter of fact it is no longer easy to break it. Something similar may also be true in the case of temporary aphasia. It may be that the categorical basis in the brain of the person's ability to construct and to recognize grammatical sentences remains undamaged, but that the mechanisms that allow this ability to be exercised are no longer operative.

In addition to the arguments above, Chomsky seems to assume that anyone holding the thesis that our grasp of grammar is a practical capacity will also be an opponent of nativism. He apparently thinks that such a thesis must go together with a picture of language-learning as a mere matter of *habituation*, of developing a skill through repetitive activity. But there is no reason why these ideas should be connected. We could agree with Chomsky's arguments for the innateness of the language faculty (as indeed I have). We could also agree that the categorical basis of a mature speaker's grasp of their language must be structured in a way that reflects the systems of rules developed by linguists. For how otherwise are we to explain the *productivity* of their ability (that is, their ability to produce, and to recognize as grammatical, sentences previously unencountered)? Yet we can combine all this with a denial of Chomsky's claim that this categorical basis—structured though it may be— is appropriately described as a system of knowledge.

The motive for such a denial is as follows. Recall that knowledge implies belief. So Chomsky's claim requires speakers to be credited with non-conscious beliefs concerning the grammatical rules that govern their native language. Now there is no objection in principle to the idea of non-conscious beliefs. For example, consider a cyclist who successfully negotiates obstacles in the road while their conscious attention is wholly directed elsewhere. Thus I might yesterday have been thinking so hard about the content of this chapter that I was unaware of what I was seeing while I cycled, so that if you had asked me even a moment later I should have been unable to say what I had been doing on the road. Yet an observer would say that I must have *seen* the vehicle parked by the side of the road, since I turned the handlebars to avoid it.

In a case such as this it seems right to say that I had a non-conscious belief that there was a vehicle there. But notice that this is because the explanation of my behaviour fits neatly into the practical-reasoning-model. It is because I *wanted* to reach my destination safely, and *believed* that there was a vehicle in the way, and *believed* that unless I turned the handlebars I should hit it, that I acted as I did. It seems essential to the very notion of belief (whether conscious or non-conscious) that beliefs are states that are apt to interact in a distinctive way with other states of desire and belief in such a way as to cause action. Yet it is this pattern of interaction that is apparently missing in the case of our supposed belief in the rules of grammar.

Another way of putting the point is that the structures that underlie our grasp of grammar do not have the sort of flexibility of behavioural effect that would be necessary for them to qualify as beliefs. Any genuine belief of yours can lead you to do one thing, given that you have one desire, or to do something quite different, given that you have another. If I believe that the ice on a nearby pond is thin, this may lead me to put away my skates, if my desire is safety; or to put them on, if I want to court danger; or to say to my wife 'That is a good place to skate' if I want to collect the insurance on her life. There is nothing analogous to this in connection with our mature grasp of the grammar of our native language. True enough, if we want to achieve one effect we will say one thing, and if we want to achieve something different we will say something different. But the relevant beliefs here will not be concerned with the rules of grammar, but rather that to say such-and-such in these circumstances will achieve such-and-such an effect. Our grasp of grammaticality will be equally implicated in any case, underlying in the same way whatever it is that we choose to say.

If we really did have beliefs about the detailed grammar of our language, underlying our ability to produce grammatical sentences, then one would expect that they could be deployed equally in the service of *un*grammaticality. For compare: your beliefs about the layout of a particular building (if true) are what enable you to find your way around it. But they will be equally serviceable if you wish to give the appearance of being lost, enabling you to wander from one corridor to another without ever reaching your destination. In contrast, we do not have a finely tuned ability to produce ungrammatical sentences to order. If asked to do so, we mostly have to resort to very crude methods, such as constructing a random sequence of nouns ('Lion tiger stag'). This looks much more like the difficulty a practised cyclist encounters if asked to ride their bicycle *badly*. The best that most of us can do here is to wobble a bit and then fall off. (It takes training to become a clown.)

It is not easy for those who possess a skill to behave as if it were absent, whereas it generally *is* easy for those who have a belief to behave as if they lacked it. In this respect our grasp of the rules of grammar seems much more like a practical ability than a set of beliefs. I therefore conclude that, while Chomsky is correct in claiming that much of our grasp of language is innate, he is wrong when he says that the mental structures that constitute that grasp should be counted as innate knowledge. (Indeed, he may even be wrong in thinking that grammatical rules are explicitly represented in the mature structure of the language faculty. Rather, they may be implicit in its mode of organization, in such a way that the system behaves in accordance with those rules, but without actually consulting them.[7])

Structure in Vision

There is a case for thinking that many other faculties besides language have innate constituent structures, embodying information about the world. Here I can only gesture at some of a rich body of evidence concerning the faculty of vision in particular.

First, it appears to be innately determined that we should perceive the world three-dimensionally. (This may be why even phenomenalists, who have become convinced that there in fact exists no independent reality, continue to *see* objects as distributed in three-dimensional space around them.) One argument for this conclusion is that it is hard to imagine how we might learn, employing only general learning principles, to interpret the patterns in a two-dimensional visual field as representing objects in three-dimensional space. There are of course cues of various sorts, of the kind exploited by landscape painters, such as perspective and shading. But it is hard to see how a child who as yet had no idea of distance could be supposed to deduce that these were indeed cues to three-dimensionality.

It might be objected that this argument is too swift. For it could be maintained that a child will learn to interpret two-dimensional images in three-dimensional terms by moving

---

[7] For discussion of this issue, see the papers by C. Peacocke and M. Davies in A. George (ed.), *Reflections on Chomsky* (Oxford: Blackwell, 1990).

around in the world, and noticing how the two-dimensional image changes.[8] In fact, however, three-dimensional vision seems to be in place *before* human babies are capable of independent motion. At least, they will attempt to move away from a visual cliff (that is, an apparent drop or drop rendered safe behind glass) as soon as they are old enough to control their movements. Moreover, it has been shown that three-dimensional vision is innate in at least some other species. In particular, new-born chicks will stay away from a visual cliff, despite having had no previous visual experience. Considerations of simplicity then suggest that three-dimensionality is innate to human vision also.

Additionally, the development of computational theories of vision has made it seem increasingly likely that seeing in three dimensions (as well as much else in our visual faculty) is innately determined.[9] For as scientists have begun to construct computer models of how the visual faculty can work out the details of the visual scene from the two-dimensional information embodied in the pattern of activation of the rods and cones in the retina, it has became obvious that these computations simply could not be carried out realistically without a rich background of innate assumptions. Amongst these assumptions is that of three-dimensionality. Other suggested assumptions, whose innateness would make the operation of our visual faculties a good deal more intelligible, are that most moving bodies are rigid, and that most perceived change (once the movements of our own head and eyes have been allowed for) results from objects moving from one place to another. These have then received dramatic confirmation in experimental findings.[10]

Suppose, for example, that you are looking at a television screen on which a circle is displayed. The circle then collapses, through narrowing ellipses, into a straight line, and then expands up to a circle again. What you will in fact *see* is a circle rotating in three-dimensional space (the straight line being the

---

[8] Berkeley develops a theory of this sort. See his *Principles*, sects. 42–4.

[9] The ground-breaking work on this topic is David Marr, *Vision* (San Francisco: Freeman, 1982).

[10] Chomsky details many of these findings in *Language and Problems of Knowledge*, ch. 5.

case where you see the circle end-on). Your visual faculty seems to embody the information that most changes of this sort do indeed result from the motion of rigid bodies, rather than from the (equally possible) collapse and expansion of bodies that are flexible.

In another clear demonstration of the importance of a rigidity assumption, experimenters attached tiny points of light to the main human joints in an otherwise completely darkened room. Observers were nevertheless able to interpret the movement of the lights as a person walking. If the lights were positioned in the middle of the rigid portions of the limbs, on the other hand, then subjects could make no sense of the resulting movements. Apparently the visual system is structured in such a way as to assume that the lights are linked by rigid bodies (as they are in the first experiment, but not in the second).

There have also been a number of dramatic experiments conducted involving apparent motion. If a light is displayed at one point on a television screen, which is then extinguished and replaced by a light at a different point, what you will in fact see (if the timing of the change is correct) is the first light moving to the position of the second. Indeed, if a bar is positioned between the two lights, what you will see is the first light moving around the bar to the position of the second. And if the first light is red and the second blue, what you will see is the light change colour while it moves. What seems to be happening is that your visual faculty automatically assumes that if changes take place in what you see, then this will normally be the product of real motion, interpreting perceived change in just such a manner if possible.

It is easy to understand from an evolutionary perspective why the basic structure of our visual faculty should be innate. All human beings, at whatever times and in whatever places, have inhabited a world that is basically the same, at least in so far as it consists of a range of differently shaped, middle-sized, mostly rigid, moveable objects against an immobile three-dimensional background. There are then great advantages to the individual, in terms of both time and reliability, if the appropriate representational system is innately structured. For in that case the individual does not have to waste time learning such a system, and there is no risk of the inevitable mistakes that would occur in the course of such learning. Both of these facts would appear to have considerable survival value, making the innateness hypothesis likely.

In conclusion, it would seem that the classical empiricists were radically mistaken in denying that our various mental faculties embody innate information about the world. Not only does the language faculty contain information about human languages, but the visual faculty contains information about the objects in space around us. And something similar is probably true in connection with other faculties as well. Yet the innate structures of these faculties, while giving rise to innate knowledge of a rather particularized sort (such as your knowledge that a given sentence is ill formed), should not in themselves be counted as knowledge.

7

The Case For Innate Concepts

In this chapter I shall consider the arguments for believing in innate concepts, leaving discussion of innate knowledge to chapter 8. (Recall from chapter 4 that concepts might be innate even though no knowledge is, and conversely that some knowledge might be innate even though no concepts are.)

Kinds of Concept Possession

We can distinguish at least three different senses in which a creature might be said to possess a given concept. First, this might be said if the organism has an appropriate discriminatory capacity, being able to act differently depending upon whether or not instances of that concept are present. For example, pigeons can learn to distinguish triangles from other shapes (pecking at triangles, but not squares or circles, to gain a reward). They are sometimes described in consequence as possessing the concept of a triangle. Indeed, pigeons are capable of learning perceptual discriminations that have a remarkable degree of sophistication. They can be taught to peck at a photograph for food only if it contains some sort of representation of a human being, whether in the foreground or background, sitting or standing, profile or full-face. They can do the same for photographs of water or trees, or indeed for photographs of a particular person.[1] Such pigeons might be said to possess the concept *human appearance*, or the concept *water appearance*.

It is doubtful, however, if this notion of a concept takes us very far. In particular, it is doubtful whether concepts in this sense will figure in any genuine beliefs. There are two points to be made here. The first concerns the notion of belief-content. It

[1] See Stephen Walker, *Animal Thought* (London: Routledge, 1983), ch. 7.

is that beliefs have contents that are essentially structured out of recombinable elements. The belief that grass is green contains the concepts *grass* and *green*, which can each figure in many other beliefs, such as the belief that grass is inedible or the belief that emeralds are green. Indeed, these same concepts can also occur in the content of other propositional attitudes such as hopes and desires, as when I hope that the grass will grow, or want a green shirt. The only candidates for the content of the pigeon's belief, on the other hand, would be 'There is a human appearance here' and perhaps 'Human appearances are a source of food'. There is no reason to think that the concept of a human appearance is indefinitely recombinable into other contents and other propositional attitudes.

The second point about beliefs concerns their causal role, which has already been stressed in previous chapters. It is essential to beliefs that they should be capable of interacting with desires and other beliefs in such a way as to cause behaviour, where that behaviour is then subject to the practical-reasoning-model of explanation. Yet it appears extravagant to try to explain the pigeon's behaviour by saying that it *believes* that the photograph it is pecking contains a human appearance, *wants* something to eat, and *believes* that pecking human appearances produces food. This is because the pigeon's repertoire of action is so extremely limited. The capacity to discriminate human appearances plays no other part in the pigeon's life than in the production of a pecking response. There is insufficient flexibility and variety of behaviour here for us to consider seriously applying the practical-reasoning-model of explanation.

The second possible sense in which an organism might be said to possess a concept arises naturally out of the limitations of the first. We could say that a creature possesses such-and-such a concept provided that we are prepared to ascribe to it a variety of beliefs or desires that involve that concept, where we then use the states we have ascribed in explanations of the creature's behaviour that fit the practical-reasoning-model. In this sense some of the higher mammals, at least, possess concepts, since we take seriously the attribution of beliefs and desires to them. For example, we may explain the behaviour of a dog by attributing to it the sequence of thoughts 'I want to get the ball.

The ball is on the table. If I jump on the chair I can reach the table. So I shall jump on the chair.' Moreover, the behaviour of a dog certainly exhibits a wide variety of ways in which it can interact with a ball—fetching, chewing, chasing, and catching—suggesting that the concept of a ball (or something like it) does form a component in a number of distinct canine beliefs and desires.

However, recall from chapter 6 that beliefs and desires can be non-conscious. I can *believe* that there is a large obstacle in the road, and consequently turn the handlebars of my bicycle, because I *want* to reach my destination safely and *believe* that unless I turn the handlebars I shall crash—all without conscious awareness of what I am doing or why. This then gives us our third possible sense of concept possession, where an organism may be said to possess a given concept provided that it entertains *conscious* thoughts in which that concept figures (or is at least apt to entertain such thoughts in appropriate circumstances). A sufficient condition for possessing a concept in this sense would be the ability to use correctly the corresponding term of a natural language. If you have the capacity to use appropriately, and to understand, statements in which the word 'grass' occurs, then this will be sufficient to show that you possess a conscious concept of grass. We can leave open the question whether relevant language mastery is also a *necessary* condition for conscious possession of a concept, hence leaving open the question whether any non-human animals possess concepts in our third sense. The answers to these questions will depend upon the nature of the difference between conscious and non-conscious thought, and on whether it is true that only language-users can entertain conscious thoughts.[2]

Armed now with the distinction between these three different notions of concept possession, we can see immediately that our basic repertoire of discriminatory-capacity concepts must be innate. If we could not, in the first place, respond differently to stimuli of different colours or temperatures, or to lines and boundaries within our visual field, then we could never learn anything else. These elemental capacities for discrimination must be built into the very structures of our perceptual

---

[2] On these issues see my 'Brute Experience' *Journal of Philosophy*, 1989, and *Tractarian Semantics*, ch. 10.

apparatuses. Moreover, even where the discriminations in question are somewhat less basic (and certainly acquired over time), such as the ability to tell human faces apart from one another, it may be that these discriminations are not really learned. Our face-recognising mechanism may contain much innately determined information about the forms and limits of expression of the human face. So even these may count as innate (but locally triggered) concepts.

It is also very likely that the basic conceptual components of non-conscious belief are innate. For, as we noted towards the end of chapter 6 in connection with vision in particular, it seems plausible that much of the representational structure of our perceptual faculties is innately fixed. That we see individual (often movable) objects against a relatively stable three-dimensional background is probably not something that we learn. In which case many of the various perceptual beliefs that can govern our actions without becoming conscious will count as employing innate—though perhaps locally triggered—concepts. (Are we then to allow, in the light of my previous example, that the non-conscious concept *handlebars* is innate? This seems implausible. But there will surely be some more neutral way of characterising the belief that governs my actions as I ride—for example, that turning the rigid bent structure in my hands will alter my direction of motion. Here the constituent concepts may plausibly be said to be innate.) The beliefs in question will be those that enable us unreflectively to pick up and turn over objects in our hands, to step over obstacles in our path when walking, to turn our heads in the direction of a sound, and so on.

Since the case for innate discriminatory abilities is a powerful one, as is the case for innate repertoires of non-conscious perceptual concepts, all interest devolves on to the question whether conscious concepts are innate. This is the issue on which I shall concentrate for the remainder of this chapter.

Are Concepts Taught?

It is clear that many linguistic concepts are universal, being common to all human societies and to all natural languages. But this in itself cannot provide sufficient reason for believing those concepts to be innate. Nor do we need to appeal to the hypothesis of a common historical origin to make the point, as we did in the case of universal features of grammar in chapter 6. For the universality of such concepts is better explained by their utility. Societies that did not have concepts of time and place, motion and rest, causality and knowledge, would not survive for very long, let alone prosper. Our universal concepts can be thought of as prerequisites for successful social life.

The very fact that many of our concepts have a degree of utility that makes them unavoidable does, however, provide some slight reason for believing them to be innate. For, as we argued in chapter 6, in connection with perceptual concepts of space, the fact that certain linguistic concepts can be regarded as amongst the standing conditions of all human life means that there would be survival value to the individual if those concepts were innate. This would allow valuable cognitive resources in the developing child to be directed elsewhere. It would also avoid the risk of mistakes occurring, which attends all learning. But the most that this argument shows is that it would not be entirely *surprising* if those linguistic concepts that are universal were innate; it does not in itself provide sufficient reason for believing that they are.

Some of the points that gave rise to Plato's Problem in connection with the learning of grammatical rules hold in the case of concept-learning as well. Children acquire a huge number of concepts in a short span of time. (Chomsky claims that at certain periods of childhood as many as a dozen new concepts are acquired each day.[3]) Moreover, they acquire these concepts on the basis of fragmentary and mostly positive data. For, just as in the case of grammar, very little explicit teaching of concepts actually takes place. In general, adults simply *use* concept-words in the presence of children, providing them with a sample of applications that is necessarily limited, only occasionally correcting the child's own mistakes. Not only this, but much of the data is, from the child's point of view, corrupt. For much adult usage consists of irony, metaphor, jokes, and teasing, in none of which are concept-words used literally, and

[3] See his *Language and Problems of Knowledge*, 27.

which can only be understood when the literal meanings of those words are already known.

However, we cannot assess how powerful the analogue of Plato's Problem is in connection with concept acquisition until we have a better idea of what concepts themselves are. For much of the burden of the argument for the innateness of grammar was taken up by the claim that the grammatical rules governing natural languages form a hugely subtle and complex system. If, in contrast, concepts were to turn out to be relatively simple entities, then the argument for their innateness would be correspondingly weaker. To this issue I shall shortly return.

Fodor on Concepts

Jerry Fodor has recently provided arguments for the innateness of at least many of our conscious concepts.[4] The first of these arguments arises out of an objection to empiricist theories of concept acquisition, and is similar to the one we used against Locke's account of abstraction in chapter 4. Fodor claims that the empiricist view must be that concepts are learned, rather than merely triggered, and then points out that the only theory of learning that has ever been seriously developed involves hypothesis formation and testing. In order to *learn* the meaning of the word 'cat', for example, a child would have to do something like the following. It might first form the hypothesis that 'cat' means 'animal', adjusting this as further data becomes available (such as a parent saying 'Look at that dog' when they pass a dog in the street), until finally it settles on the hypothesis that 'cat' refers to cats. Then the obvious point is that in order to go through such a procedure, the child must already have available to it the concepts *cat* and *animal*. So if this were the only story of concept acquisition that an empiricist could tell, it would follow immediately that a large repertoire of concepts must already be innate before a child could begin to learn the meanings of the words in its native language.

A natural objection to this argument is that it assumes that possession of a linguistic concept is a genuine item of propositional knowledge—a matter of knowing *that* a given word

[4] See his *Representations*, ch. 10.

means such-and-such. But it would be equally plausible (if not more so) to maintain that possession of such a concept is a practical capacity—a matter of knowing *how* to use a given word. On this view, to possess a concept is to be capable of classifying presented items correctly, as well as having the ability to employ the appropriate word in a variety of sentence-constructions. If this is right, then acquiring linguistic concepts will be much more like acquiring a practical skill, such as the ability to ride a bicycle, than it is like acquiring a new item of information. And no one would want to claim that bicycle-riding is an innate ability (although, of course, it does make use of our innate capacities, such as those of balance and leg movement).

There are a number of replies to this objection available to Fodor. The first is that acquiring linguistic concepts cannot be a mere matter of habituation, or repetitive practice. (This parallels Chomsky's point about the acquisition of grammatical rules, discussed in chapter 6.) For our concept-words admit of an indefinite number of possible uses, and a competent speaker will be able to understand combinations of concepts that they have had no previous experience of. Someone who understands the word 'cat', for example, does not merely have the ability to say 'cat' when a cat is present, but also to *ask* whether there is a cat present, to assert that there is *no* cat present, and generally to connect that word together in an indefinite number of ways with the other words of the language that they understand. So even if we grant that possession of a concept consists in a practical capacity, rather than being a matter of genuine propositional knowledge, still it must be a special sort of capacity, whose categorical basis in the brain somehow reflects the structure of the concept, and its connections with other concepts. And then acquiring that concept cannot be merely a matter of practising an activity, like learning to ride a bicycle.

Fodor would also reply that in any case all cognitive processes take place in what he calls 'a language of thought'. This would include the process of acquiring the capacities that we are supposing to constitute grasp of linguistic concepts. Fodor's view is that all cognitive processes are *computational*, in the sense that they involve operations upon sentences or sentence-like constructions. He holds that this is so whether the processes in question are conscious or non-conscious, or whether they involve the organism as a whole or only subpersonal systems such as vision or the details of language comprehension. So even if possession of a concept is best thought of as a kind of practical capacity, rather than an item of propositional knowledge, the process of acquiring that capacity will still involve hypothesis formation and testing in the language of thought. Then, since all processes, including those involved in learning, take place in the language of thought, this language itself must be innate. It then follows that some concepts—namely the concepts of the language of thought—must be innate in order that linguistic concepts may be acquired.

The idea of a language of thought is controversial. But, even granting it, Fodor's argument is less powerful than it seems. For the most that it does is to force us to recognize that *non*-conscious belief-concepts are innate, in order that conscious (linguistic) concepts may be acquired. But we had already granted that in any case. And it does not follow that conscious concepts are themselves innate. Indeed, the appropriate conclusion may be weaker still. For it would be much more plausible to assimilate the concepts involved in cognitive processing generally to the discriminatory-capacity concepts discussed earlier. For it is doubtful whether cognitive models of hypothesis formation and testing within some subpersonal mental faculty (such as that of language processing) will genuinely fit the practical-reasoning-model of explanation. So it may only follow (if we accept the idea of the language of thought) that we are committed to the innateness of subpersonal discriminatory capacities. But again, this was something that we had accepted in any case. The interesting question is whether conscious concepts (or something like them) are innate.

Fodor on Definitions

Fodor does have a different argument for the innateness of many conscious concepts, which turns on his claim that most linguistic concepts are in fact *indefinable*. For he argues that the main point of contention between empiricists and nativists about concepts concerns the range of concepts that are definable in terms of a primitive conceptual basis. According to Fodor, the empiricist maintains that all other concepts may be defined in terms of a set of innate concepts that are *sensory*, that arise directly out of experience (the process of construction through definition then providing the empiricists with their theory of learning). The nativist maintains, on the contrary, that most of our concepts are triggered rather than learned.

There is one way in which Fodor definitely goes wrong here. This is in his claim that an empiricist must count the basic sensory concepts as innate (triggered by experience, rather than learned from it) because they are not acquired via a process of hypothesis formation and testing. For this account of learning is too narrow. Empiricists such as Locke and Hume certainly thought that our basic concepts are learned from experience, by virtue of their being a record of the initial experience laid down in memory. What Fodor overlooks is that our common-sense concept of learning divides into at least three distinct kinds. There is learning by means of an inference to the best explanation, which is what forms Fodor's paradigm. But there is also learning how to do something, such as ride a bicycle. And there is memory-based learning, as would occur when one comes to know what someone looks like, by meeting them. In fact the early empiricists proposed that basic sensory concepts are learned in the third of these modes, on the basis of prior acquaintance. All the same, Fodor is quite right about the main issue, which is that it will count in favour of the rationalist thesis that most of our concepts are innate, if it turns out that they cannot be defined at all, let alone in purely sensory terms.

Fodor's main argument for indefinability is based upon our failure, historically speaking, to come up with agreed definitions. Despite generations of co-operative labour by analytic philosophers, there is hardly a single concept whose analysis is generally accepted. The best explanation of this phenomenon, Fodor thinks, is simply that there do not exist any definitions to be found. But in fact an alternative explanation is suggested by Chomksy.[5] This is that most of our concepts are extremely complex, being related to one another in all sorts of subtle ways. This seems on the face of it to be equally acceptable. However, an interesting possibility opens up here. For it may be that in

[5] See *Language and Problems of Knowledge*, ch. 1.

either case we shall have an argument for nativism. For Chomsky, in effect, turns this alternative explanation of failures of definition into the analogue, in the case of concepts, of Plato's Problem for grammar acquisition. Supposing that the system of concepts of a mature speaker is very subtle and complex, then the problem of how a child manages to acquire this system on such a slender basis really does become pressing.

But in fact, even supposing that Fodor is right that most concepts are indefinable, this need not commit us to nativism. For, as he himself notes, 'indefinable' does not necessarily imply 'unstructured'. It may be that complex concepts are built up out of simpler ones, without being definable in terms of them (at least if the definitions are required to take the form of statements of necessary and sufficient conditions of application). Just such a possibility will be realized if concepts are represented in the mind by *prototypes*, which is a thesis for which much psychological evidence has recently begun to emerge.

Concepts as Prototypes

A prototype is a specification of a set of prototypical properties, taken together with a weighted similarity measure. The prototype for the concept *dog*, for example, would include such features as *barks when angry*, *wags its tail when happy*, *is a mammal*, *has dogs as parents*, *eats bones*, and so on. But there is no suggestion that all dogs must necessarily have all these features. Rather, deciding whether something is a dog is a matter of judging whether it is sufficiently similar to the prototypical dog.

Fodor considers this theory, but thinks he has a swift dismissal of it. He claims that it cannot provide an account of compositionality—of how we can combine together linguistic concepts to form new ones.[6] He asks, for example, how the meaning of 'brown cow' is supposed to be determined from the prototypes of 'brown' and 'cow'. But the reply is surely straightforward. It is that the meaning of 'brown cow' may be given by 'is sufficiently similar to this (the prototype of brown) *and* is sufficiently similar to this (the prototype of cow)'. The point is that the meanings of complex concepts formed by

---

[6] *Representations*, 296–8.

putting together individual words need not themselves be prototypes. The false assumption in Fodor's argument is that if prototype theory is correct, then the concepts expressed by lexically complex items such as 'brown cow' must themselves be prototypes. Rather, such concepts may be logical constructions out of prototypes—building them up using notions like *and* and *or*.

However, if concepts like *brown cow* are logical constructions out of prototypes, where do these logical notions come from in their turn? I think that they could very plausibly be allowed by an empiricist to be innate. They can be regarded as belonging to the given (innate) logical structure of the mind. For these notions do not in themselves constrain what the world may be like. Empiricists should only object to the innateness of concepts that seem to carry information about the world—whereas the logical concepts will be equally applicable no matter what our experience of the world may be like. So Fodor's argument against the empiricist fails, even given the way in which he sets up the debate. It may be that most of our complex concepts are either prototypes or logical constructions out of prototypes, where these in turn are constructed out of simple ideas derived from experience (but without being definable in terms of those ideas).

I shall not now review the psychological evidence supporting the thesis that many of our concepts have prototype structure.[7] Rather, supposing that such a thesis is correct, let us ask whether it is sufficient to defend empiricist theories of concept acquisition. One point can be made straight away, in the light of what was said in chapter 4. For it would be just as implausible to maintain that all complex concepts can be *constructed* out of a basic set of purely sensory ones as it would be to claim that they can all be *defined* in such terms. (For example, try giving a prototypical set of sensory concepts that would even begin to be adequate to express the concept of causation.) So if empiricism were committed to reducing all concepts to those that describe our private sensations, then the introduction of prototypes would bring it no particular advantage.

However, recall from our earlier discussion that we are allowing there to be an innate set of perceptual belief-concepts

[7] On this, see George Lakoff, *Women, Fire, and Dangerous Things* (Chicago: Chicago University Press, 1987).

that, so far from being purely sensory, involve representations of physical objects in three-dimensional space. This might provide a sufficient basis on which to construct complex prototypes. For example, a child might first acquire, through experience, a perceptual *paradigm* for a given concept. On hearing an object described as a cat for the first time, it lays down a representation of that particular cat in memory. (It is this stage that presupposes innate perceptual belief-concepts.) Then, through further experience, it begins to acquire a similarity measure for judging whether a given object is sufficiently close to the paradigm to be a cat. For example, the child might start by overextending the concept, judging that a dog is sufficiently close to the paradigm, until it is either corrected by its parents or hears the dog described as such. As it begins to acquire perceptual paradigms and similarity measures for other concepts, the child also starts to cross-classify, building up the full prototype for 'cat'. For example, having acquired a paradigm for 'tail', it learns that cats have tails, and having acquired a paradigm for 'mammal' it learns that cats are mammals, and so on.[8]

If this picture, or something like it, proved to be correct, then the empiricist view of the acquisition of conscious concepts would be largely vindicated. The initial perceptual paradigms would be given in experience (although requiring an innate perceptual basis), and the later cross-classifications could also be learned through experience. The similarity measures, too, might be learned from experience, if it were to turn out that parental teaching is crucially necessary to prevent children from over-extending or underextending their concepts. Indeed, it would be a virtue of this sort of account that it could overcome the problem of a historical regress, outlined in chapter 4. We need only suppose that the similarity measures were first settled upon *causally*, rather than arrived at by prior teaching or conscious choice. This could either have happened by chance, or as a result of some similarities being more salient, or more relevant to human concerns, than others.

No doubt it is a mistake to treat all concepts as if they were alike. It may be that some are prototypes, and are acquired from

---

[8] A theory of this sort is developed by Andrew Woodfield, 'On the Very Idea of Acquiring a Concept', in James Russell (ed.), *Philosophical Perspectives* (Oxford: Blackwell, 1987).

experience, and that some are not. One class of concepts for which the prototype theory looks particularly implausible are psychological ones such as *belief*, *desire* and *intention*.[9] (In fact, it is in connection with such concepts that Chomsky urges his alternative response to the lack of agreed definitions, opting for an explanation in terms of complexity, and arguing for innateness as the only solution to Plato's Problem.) These will come into focus in the next chapter, where we consider the arguments for saying that our knowledge of our own psychology is innate, together with its constituent psychological concepts. For the moment our conclusion must be that the case for the innateness of conscious concepts is largely unproven.

The Concept of Best Explanation

There is, however, one cluster of concepts that are very likely to be innate. These are the concepts involved in the appraisal of non-deductive modes of argument, particularly the concept of the best available explanation of a given phenomenon. This concept is undoubtedly a conscious (because linguistic) one, but it is notoriously difficult to define. Yet there is a remarkable degree of agreement amongst speakers in particular cases as to whether one explanation is or is not better than another. Moreover, while the concept does to some extent display prototype structure (a good explanation tends to be simple and consistent, cohere well with surrounding beliefs, have broad scope, be fruitful in generating new predictions, and so on), the constituent notions are no easier to define in their turn; nor is it easy to see how they might be derived from experience. Then, since children receive no explicit training in the use of this concept, we cannot explain how we nevertheless manage to acquire it, unless we suppose that the concept is innate— triggered by the course of our experience, rather than learned from it.

It is worth stressing that while the concept of best explanation does figure prominently in science, it is by no means an exclusively scientific idea. On the contrary, we each of us constantly make at least tacit use of it in the course of our daily

[9] Another would be the concept of *cause*. Indeed, as we noted in chapter 4, there may be a powerful case for saying that this concept, at least, is innate.

lives. For example, when I enter a lecture-hall full of students, and consequently form the belief that they have come to listen to me speak, this is because this belief provides, in the circumstances, overwhelmingly the best available explanation for their presence. Scientists probably make use of the very same concept, differing only in that their search is for the general laws or principles underlying the observable phenomena to be explained.

One cannot mention non-deductive modes of argument without discussing Hume. For Hume is famous for his discovery of the problem of justifying induction, and for his naturalistic explanation of our use of it. His view is that induction has no rational basis, either in reason or in experience. Its reliability cannot be demonstrated a priori. Nor can it be shown to be reliable by past experience without circularity.[10] Our reliance upon induction is rather to be explained by appeal to the basic principles of human psychology, particularly the laws of association amongst ideas. Hume's theory is that having seen a phenomenon repeated, we are habituated to expect it to continue in the future. But this is not to ascribe to us an innate concept. For the psychological principles that give rise to induction are supposed to be general ones. To say that our psychology is such that we do, as a matter of fact, go in for induction is not to say that our minds contain an innate concept of a good inductive argument.

Quite apart from the crudity of his psychological theory, there are two things wrong with Hume's account. The first is that it ignores the *normativity* involved in our idea of best explanation. The fact is that we do not (as a mere matter of fact) reason in certain ways, as a result of our nature. We also apply standards of appraisal to such reasonings, counting some non-deductive arguments as better than others, and counting people as right or wrong in their assessments of the merits of such arguments. Yet this is one of the distinctive marks of possession of a conscious concept. For, as we noted in chapter 3, concepts are best construed by empiricists as rules of classification. This seems to be just what we have in the present instance: we have a rule for classifying some non-deductive arguments as better than others.

[10] We shall return to consider these points in greater detail in chapter 12.

Hume's second mistake is that induction is by no means the only non-deductive mode of argument. Induction is a matter of generalizing or projecting from observed regularities. Yet we also frequently reason from such regularities to the presence of some underlying phenomenon (sometimes involving unobservables, such as sub-atomic particles) that would explain it. Indeed, induction itself is best seen as a particular instance of the more general practice of inferring to the best explanation. The reason why we infer from 'All ravens so far observed have been black' to 'All ravens are black' is that the fact that all ravens *are* black is the best available explanation for why all *observed* ravens should have been black. Yet there is no question of exhibiting inference to the best explanation as flowing from the general principles of association amongst ideas. Nor is it easy to see how it might be explained in terms of general features of any more sophisticated psychological theory. Yet all Hume's reasons for thinking that induction is cognitively basic and indispensable for us now transfer to the case of inference to the best explanation. It is therefore plausible to suppose that we have an innate grasp of the concept of best explanation, which we then employ in appraising particular arguments.

It may be objected that there are wide variations across cultures concerning what explanations will be counted as better than others. How, then, can the concept of best explanation be innate to the human species? In our culture, for example, we explain illnesses in terms of such things as the activity of viruses on the body, whereas in other cultures the same illness may be explained in terms of the malignant action of a witch. But this is to miss the point that what counts as the best explanation of some phenomenon is always relative to one's background beliefs. It is these beliefs that vary across cultures. The reason why the hypothesis that an illness was caused by the action of a witch is not a good explanation for us is that we do not believe in witchcraft. It is thus plausible that all cultures employ the same general standards for selecting between competing explanations, given their background beliefs.

Note that if inference to the best explanation is a generally reliable method of forming beliefs, then it is easy to explain how our concept of it could come to be innate. For individuals will be better able to survive if they are able to attain true beliefs about the underlying processes at work in nature, which can then be harnessed and exploited, or if they can acquire knowledge of the unseen causes of observable phenomena. (Conversely, if our concept of best explanation has already been shown to be innate, then this gives us good reason to believe that such inferences are generally reliable. For a concept could not have been selected through evolution if it did not confer survival value on the individuals who possess it. Yet it is hard to see how inference to the best explanation could have survival value, unless it is indeed reliable. We shall return to this issue in chapter 12.)

I conclude that, while the case for innate concepts is largely unproven, there is at least one concept that is probably innate, namely our concept of best explanation. This result will prove to be of some importance in chapters 11 and 12, when we come to discuss the problem of scepticism.

8

The Case for Innate Knowledge

In this chapter I shall consider the most likely candidate for substantive innate knowledge, namely our beliefs about our own psychology.

Innateness and Reliability

Recall from chapter 5 that in order for there to be innate knowledge, we must possess some innate beliefs that are not only true, but caused by a reliable process. While classical rationalists held that the process in question was divine intervention (our innate beliefs having been implanted in us by a veracious God), this idea is no longer taken seriously in scientific cultures. In order for a claim to innate knowledge to be given even cursory consideration today, the proposed belief-causing process must be a *natural* one. Indeed, the only suggestion that is consistent with current scientific knowledge is that innate beliefs might be determined through evolutionary selection. For science tells us that this is the manner in which all other innate characteristics have been arrived at, whether in ourselves or in other organisms.

What reason is there for thinking that natural selection would be a reliable process, supposing that it resulted in some innate beliefs? Notice, first, that true belief has immense survival value for any organism, such as ourselves, much of whose behaviour is caused by the interaction of beliefs and desires. For in general an organism's projects will only succeed if based upon beliefs that are *true*. This is not to say, of course, that action undertaken on the basis of a true belief is guaranteed to succeed. I may set off walking through the desert, correctly believing that there is water to be found in that direction, but perish of thirst before I reach it. But often (though not always) the failure will result from the falsity of some other belief. Thus, in this example, it may be my false belief that the water is close enough for me to reach by walking that is responsible for my death. (On the other hand, I may simply have lacked any alternative course of action, so that I would have perished whatever I did.)

It is also true that action undertaken on the basis of a false belief is not guaranteed to fail. One way in which the action may nevertheless succeed is where the belief, while false, is sufficiently close to the truth. For example, although my belief about the direction of the oasis may be strictly incorrect, it may still lead me to within sighting distance of the water I need. If an action based upon a wholly false belief succeeds, however, it will only succeed by accident. For example, although walking in the wrong direction to find the oasis I am aiming for, I may be lucky enough to stumble across another, and hence survive anyway. So wholly false beliefs will not have survival value in the long run, and in evolutionary selection it is the long run that matters. What seems undeniable is that organisms (of the sort that act on beliefs) will only survive, in general and in the long run, if they base their actions on beliefs that are true, or at least close to the truth. So if any innate beliefs have arisen through natural selection, we should expect them to be at least approximately true.

An objection to this line of argument, however, is that natural selection can explain features of organisms that do *not* have survival value, provided that they are by-products of things that *do* have it. If this were true of innate beliefs, there would be no particular reason to expect the beliefs in question to be true ones. (Indeed, even if those beliefs did happen to be true they would not count as reliably acquired, and hence would not be known. For their truth would then turn out to be accidental relative to the processes involved in their acquisition.) Yet this objection cannot be sustained, at least in the absence of any concrete proposal. For it is extremely difficult to understand how innate beliefs might be a by-product of some other innate characteristic of human beings, without themselves having a value in survival contingent upon their truth.

Another point to notice is that natural selection has conferred on us belief-acquisition processes that are generally reliable.

Beliefs arising from perception and memory are, on the whole, true. (Those sceptical of this will be answered in chapters 11 and 12.) So even if evolution does not issue directly in any beliefs, it has resulted in innate mechanisms for acquiring beliefs that are fairly reliable. It is therefore, to that extent, a reliable process in its turn. Moreover part, at least, of the inaccuracy of our perceptual mechanisms can be accounted for in terms of a competing constraint, which would not operate (or not to the same extent) in the case of innate beliefs. This is speed. Perceptual mechanisms need to be fast as well as reliable. So there may be survival value for an organism in some sacrifice of reliability, if this is more than offset by a corresponding gain in speed. For example, it will be an advantage to have a perceptual mechanism that informs me extremely quickly that there is a predator running towards me, even if—as a result of the speed with which it operates—the mechanism often informs me that there is a predator approaching when there is not. Better sometimes to flee unnecessarily than never to have time to flee at all! But in connection with innate beliefs, as opposed to perceptual mechanisms, there is presumably no demand for speed (except perhaps speed in accessing and calculating with those beliefs). So evolutionary selection of innate beliefs would, if anything, be even *more* reliable than that of perceptual mechanisms.

It is possible to imagine cases where an innate *false* belief would be an aid to survival. For example, an innate belief in the magical properties of a particular plant, which in fact contains a powerful medicine, might prove very useful to those who live in the region where that plant flourishes. But such cases are rendered unlikely when one remembers that in order to have been selected through evolution, a belief would have to prove useful over a time-span that is extremely long in comparison to human history, and in a wide variety of differing circumstances. I therefore conclude that if evolution has resulted in any innate beliefs, then those beliefs will very likely constitute innate knowledge.

Recall that in chapter 6 we discovered some candidates for innate knowledge, in the course of our discussion of innate mental structure. Thus, while we disagreed with Chomsky's claim that the innate structure of the language faculty should be counted as genuine knowledge, we noted that our conscious beliefs about whether a particular sentence is or is not well formed will count as innately known. For such beliefs are true, and are caused by a reliable process, provided that they result in a systematic way from the underlying structure of the language faculty. But if this is the total extent of our innate knowledge, it will be of very marginal significance. For it is not these beliefs themselves that explain our capacity to formulate and to recognize grammatical sentences. They rather arise out of what does explain that capacity, namely the innate structure of the language faculty, together with the subject's parameter-setting experience of their native language. Indeed, it is arguable that the very idea of a sentence being well formed or ill formed only arises in cultures where for political reasons attempts are made to standardize different dialects. I am told, for example, that linguists working in the field often have considerable difficulty in explaining to the members of the language community they are studying just what it is they are after when they seek the speakers' judgements as to whether or not a given native sentence is well formed.

In chapter 6 we also allowed that other faculties, such as vision, are innately structured, embodying information about the world. If this is so, then it seems that perceptual beliefs will (somewhat strangely) count as innate, being triggered by experience rather than learned from it. But perhaps a better way to put the point is this. The innate structure of the visual faculty may give us innate (locally triggered) knowledge that there are, in general, physical objects distributed in three-dimensional space around us. But then our particular perceptual beliefs—for example, that there is a cylindrical object resting on a flat surface in front of me now—are based upon experience. For it is only our general ability to perceive in three dimensions that is unlearned. On that basis we then *learn* where particular objects are by seeing. So our detailed perceptual knowledge of the world around us should not be counted as innate.

If we are to show that we have innate substantive knowledge of detailed truths about the world, we need to consider some other candidates. One example may be provided by our knowledge of our own psychology.

Folk-Psychology

We have an immense network of common-sense beliefs about the mind. These beliefs concern the relationships of mental states to one another, to the environment and states of the body, and to behaviour. This collection of beliefs is generally referred to by philosophers today as 'folk-psychology'. It includes such beliefs as these: that pains tend to be caused by injury, and tend to prevent you from concentrating upon other tasks; that perceptions are generally caused by the appropriate state of the environment (for example, the experience of a tiger being caused by the presence of a tiger), and are often laid down in memory; that if you want something, and believe that performing an action or sequence of actions will enable you to get it, then you will normally (other things being equal) do that thing; that decisions are often the result of prior deliberation, and generally lead you to perform the action decided upon; that when you assert something, you generally believe it, and when you say you are going to do something, you generally do it; and so on.

It is important to stress that folk-psychology should only be understood to cover the set of common-sense generalizations about the human mind that hold good independently of context or culture. For it is perfectly obvious that our *particular* beliefs about someone's psychology—concerning their individual desires, intentions, and beliefs—will not be innate, but will rather be learned from observation of their behaviour. Similarly, when we arrive for the first time in a foreign country we may not be able to take much for granted about their customs, or about what they believe or value. So we shall not be able to rely upon such culture-relative generalizations as that a hand-shake will be treated as a sign of greeting. But what we *can* take for granted is that the general way in which their minds work will be similar to ours—for example, that they will generally believe what they see. And it is this general knowledge that we shall largely rely on in interpreting their behaviour, and in beginning to construct detailed pictures of their psychology.

There are a number of points to notice about folk-psychology. The first is that it is extremely complex, consisting of perhaps many thousands of distinct generalizations. This complexity can

easily pass unnoticed, since we are so at home within folk-
psychology that we are barely aware of its existence. But the
complexity emerges as soon as we try to articulate everything
that common-sense tells us about the mind. (To see this, just try
completing the list of generalizations started in the paragraph
before last!) We might be tempted to say that folk-psychology is
no less complicated than the human mind itself. But this would
be an exaggeration. For a great deal of the operation of our
minds is, no doubt, non-conscious; whereas folk-psychology is
mostly concerned with our conscious mental life. But even this
is complicated enough.

The second point to notice about folk-psychology is that,
while it may be loosely structured and imprecise, it is also
remarkably successful.[1] Because of it, the actions of other people
as well as ourselves are often predictable, and almost always
intelligible. (We can make sense of one another, in a way that
we are able to understand hardly anything else in the natural
world.) Indeed, it is only because of folk-psychology that social
life is possible at all. We constantly rely upon it in interpreting
the utterances of other people; in recovering from their
utterances, circumstances, and behaviour a description of their
beliefs and intentions; in predicting what may be expected of
people with those beliefs and intentions in a given situation; and
in predicting the effects on other people of our own utterances
and actions. To see how successful we are in all this, reflect
upon the complexity and diversity of modern societies, and yet
on the extent to which we nevertheless manage to co-ordinate
our behaviour.

Just as it is easy to overlook the complexity of folk-
psychology, so it is easy to underestimate its success. For our
occasional muddles and misunderstandings tend to have far
greater salience for us. This is because many of the cases that
matter to us most are those where understanding is hardest to
come by. Thus one person's religious, moral, or political beliefs
may seem completely mysterious to another. Indeed, by
generalizing from and romanticizing such cases, some people
are inclined to picture human beings as wholly opaque to one
another, beyond mutual knowledge or comprehension. What

---

[1] This point is made convincingly by Jerry Fodor, *Psychosemantics* (Cambridge,
Mass.: MIT Press), ch. 1, as are many of the points that follow.

they overlook are the myriad cases of mundane success. Even those with widely differing religious beliefs can co-operate successfully in practical projects, such as building a wall together, or co-ordinating a meeting in a strange city. And many of their everyday actions will be mutually intelligible.

The third point to notice about folk-psychology, besides its complexity and remarkable success, is that it is also *deep*. As a first approximation, you might say that this depth consists in the fact that folk-psychology explains and predicts human behaviour through the interaction of *unobservables*—beliefs, desires, feelings, and thoughts. In this respect, at least, it might be compared to highly developed sciences such as physics and chemistry, which similarly explain the observable behaviour of physical substances in terms of the unobservable interactions of their parts. However, this way of putting the point presupposes a distinction between theory and observation that is highly contentious. Many philosophers would want to insist, on the contrary, that we can literally *see* someone's pain or desire, against the background of our beliefs about their circumstances and other mental states. Similarly, many philosophers of science would want to say that a physicist in the course of an experiment may—in the light of their background theoretical beliefs—be said to *see* electrons being discharged from the substance under study. These philosophers would maintain that, since what we see is always to some extent a function of what we believe, there is no motive for denying that entities such as pains and electrons are observable.

A better way to characterise the depth of folk-psychology is to stress its implicit *realism*. It commits us to the real existence of mental states and events, and real causal interactions between them. Indeed, it embodies a complex theory about the inner structure of the mind. Consider, for example, a folk-psycho-logical explanation of someone's decision in terms of the train of reasoning that led up to it. This postulates a causal sequence of real events, which were unobservable in fact, even if not in principle.

One question left outstanding is to what extent the generaliza-tions of folk-psychology are consciously believed. It might be suggested that they should rather be assimilated to the sort of non-conscious mental structures that underlie our grasp of grammatical rules. Now it may be that there are such structures here too, but it is surely undeniable that generalizations of the sort mentioned earlier will be consciously endorsed. This need not mean that they often (or in fact ever) *surface* in consciousness. For being truistic (too obvious to mention), they mostly pass unthought of. But they are certainly *available* to consciousness, since speakers will immediately assent to them if asked.

This is not to say, however, that we cannot make mistakes when we try to articulate the principles we are taking for granted. For example, at one point Fodor says that people will (other things being equal) do whatever they believe to be *necessary* to fulfil their desires.[2] But this is false if taken quite generally. What really matters is that the act be *sufficient*, or be part of a sequence of actions that is believed to be sufficient. Thus even if I would like to become a famous concert pianist, and believe that it is necessary that I should first learn to play the piano, it does not follow that I will make the attempt. For I might also believe that even if I learned the piano I should never be good enough to be famous.

Is Folk-Psychology Learned?

Folk-psychology is apparently universal to all human communities. So far as I am aware, there are no societies that fail to have notions corresponding to those of pain and anger, and that do not employ the practical-reasoning-model for explaining actions in terms of the interaction of beliefs and desires. This is, if true, a striking fact. There is hardly any other comparable body of beliefs that is equally universal. Indeed, the only real candidate is folk-physics, which includes such truisms as that it takes more energy to lift a heavy object than a light one. A case can be made for saying that this, too, is innate. But I shall not pursue the matter here.

Stephen Stich has raised doubts about the universality of folk-psychology,[3] citing the work of the anthropologist Rodney

---

[2] See *Psychosemantics*, 2.
[3] See *From Folk Psychology to Cognitive Science* (Cambridge, Mass.: MIT Press, 1983), 217–19.

Needham.[4] But in reality the anthropological evidence presented by Needham *supports* the universality of folk-psychology. What seems to count against it is only a faulty (Wittgensteinian) analysis of folk-psychological notions.[5] On such an account, these notions are conceptually tied to specific types of behavioural criteria, and are necessarily embedded in particular linguistic and cultural practices. In which case, since in many societies these practices are absent, the notion of belief, for example, may be held to be without application. But in fact, as we have seen, folk-psychology postulates a network of causally related real internal states and events, where such states will only issue in a given type of behaviour if other things are equal —that is, if the surrounding states in the network remain the same. From this perspective, it is only to be expected that people in different cultures, while enjoying the same *types* of mental state (beliefs, desires, and so on), might engage in very different behaviours, depending upon the particular beliefs and values that are current amongst their members.

However, the universality of folk-psychology is not in itself sufficient to show that folk-psychological beliefs are innate. For supposing that those beliefs are largely true, then it is clear that no human society would last very long without them. Since a shared common-sense psychology is the basis on which all social co-operation and communication rest, groups of humans who did not possess it would be hard put to survive, let alone flourish. But, for all that, it may be that the beliefs in question are empirically acquired. It may be that they were gradually discovered by emergingly successful social groups in the past, and are now passed on between generations by teaching. Or it may be that they are rediscovered by each individual in the course of normal psychological development.

In fact, however (as in the cases of grammatical rules and concepts), very little, if any, explicit teaching of folk-psychology takes place. Adults simply *use* psychological notions in the presence of children, leaving the latter to acquire by themselves the generalizations within which those notions are embedded. Indeed, Plato's Problem arises here in particularly stark form.

[4] Esp. his book *Belief, Language and Experience* (Oxford: Blackwell, 1972).
[5] See Ludwig Wittgenstein, *Philosophical Investigations* (Oxford: Blackwell, 1953).

For the body of knowledge that the child has to acquire is not only large, but structured in a very complex way. Since almost every mental state can interact with any other, folk-psychology must consist of literally thousands of generalizations. Yet recent evidence suggests that children have an adequate mastery of a great deal of it, at least, by the end of their fourth year.[6] (Indeed, since much of the evidence relates to the child's *verbal* ability, the actual time of acquisition may be earlier still. For it is often the case that understanding precedes the ability to articulate.)

Moreover, while the data available to the child may not actually be corrupt, it is certainly fragmentary and incomplete. Since much of folk-psychology is truistic (to adults), it will hardly ever be cited explicitly in explanations. No adult ever says things like 'Daddy has gone to the shops because he wants food to eat tonight, believes that there is nothing suitable in the house, believes that the shops are the best place to get food, believes that now is a suitable time to go to the shops' and so on. One simply says 'Daddy has gone shopping to get supper'. Not even the endless 'Why?' questions of a 2-year-old will lead one to articulate a generalization like 'When people want something, and believe that they can do something to get it, they tend (other things being equal) to do that thing'. So how are we supposed to imagine that the child can gather, purely by inference from its own observations, the full repertoire of psychological generalizations? Moreover, much of what is needed to formulate these generalizations is in any case hidden from the child. For, being inner, it is not available to pre-theoretical observation.

It might be suggested that the child can learn folk-psychological generalizations from its own case, by introspection. But this is highly implausible, for at least two reasons. The first is that even if we suppose that all mental states are transparently available to consciousness, as Cartesian models of the mind maintain, the relevant causal connections are not. So we would have to suppose that the young child has the ability to construct (albeit non-consciously) an explanatory theory of remarkable sophistication and complexity. It would also be very surprising that all children should happen to hit upon the *same* theory. But

---

[6] See e.g. Henry Wellman, *The Child's Theory of the Mind* (Cambridge, Mass.: MIT Press, 1990).

the second reason why introspection cannot be the source of the child's knowledge of folk-psychology is simple and devastating. It is that many mental states are not in any case immediately available to introspection. While it is true that many mental states characteristically give rise to a belief in their own existence, that they do so is itself one of the generalizations of folk-psychology that the child needs to learn. Introspection cannot, for example, show you that you went to the shops because you wanted to buy food and believed that it was a suitable time to go, if your belief in the suitability of the time did not surface in a conscious thought prior to your action. You may now, looking back, say that of course you had such a belief underlying what you did; but then it cannot be introspection that tells you that such retrospective explanation is reliable.

I conclude that Plato's Problem concerning the child's acquisition of psychological generalizations cannot be solved, unless we suppose that much of folk-psychology is already innate, triggered locally by the child's experience of itself and others, rather than learned.

Supposing that the generalizations of folk-psychology are largely true (or at least close enough to the truth to be successful), we can construct an additional argument for supposing them to be innate. For their innateness would confer considerable advantages in survival. Recall that it is our shared belief in folk-psychology that makes co-operation and communication possible. So if folk-psychology had to be learned, it would have to be learned first. No co-operation could take place between parents and children, and no other information could be acquired by children from their parents, until the relevant generalizations had been learned, and learned correctly. This would waste time and cognitive resources, at a stage in development when every additional fact that a child can learn from adults can save its life. ('Don't play with snakes; if you see a tiger, then run; don't try to climb into the cooking pot', and so on.) On the other hand, if the relevant generalizations are innate, only needing to be triggered into existence by the child's early experience, then the child can immediately embark on the important task of learning the accumulated wisdom of its society. So the innateness of folk-psychology is just what one might have expected evolutionary selection to deliver. Creatures, such as ourselves, whose survival depends crucially upon mutual knowledge of one another's psychology, will survive best if that knowledge is innate.

Indeed, recent evidence suggests that it was the evolution of a distinctively *social* intelligence that gave the primary impetus to the evolution of human intelligence in general.[7] Studies of the behaviour of our closest animal cousins in the wild, chimpanzees, show convincingly that they, too, have at least a primitive model of their fellows' psychology. The suggestion then made is that once this model is innately given, new modes of social interaction (co-operation as well as competition) become possible. The advantages in survival of such interaction would be so decisive that one would expect the model to have become rapidly more sophisticated, proliferating in the end into full-blown human intelligence. This story is certainly plausible enough to support further the suggestion that human folk-psychology is innate.

Is Folk-Psychology Correct?

Recall that in order to qualify as knowledge, a belief must at least be *true*. So if folk-psychology is to be not only innate, but innately known, it must constitute a correct theory of the mind. So far we have been assuming that it is indeed correct, but many have denied this.[8] They have maintained that there is no more reason to believe folk-psychology to be correct than there is to believe in the correctness of folk-medicine or folk-religion. But in that case it is hard to see why folk-psychology should be so staggeringly successful. Of course it is not *complete*, since there are many puzzling facts about our psychology that it cannot explain (for example, various forms of mental illness). But its degree of success within its domain (namely, normal conscious mental life) can be measured by the success of the species as a whole. For it is folk-psychology, as we have said, that underlies our ability both to cooperate and to communicate information

---

[7] See the papers collected in R. Byrne and A. Whiten (eds.), *Machiavellian Intelligence* (Oxford: Oxford University Press, 1988).

[8] See e.g. Stich, *From Folk Psychology*.

which in turn are the very foundations of human society and human progress.

It might be said in reply that human communities have flourished and prospered despite their members' having had many false beliefs—for example, beliefs in magic and witchcraft. But the difference between such beliefs and those of folk-psychology is that the latter are implicated in almost every aspect of our practical lives. Beliefs in magic, in contrast, while perhaps being highly valued by those who possess them, are relatively peripheral. It is not surprising that societies should flourish despite their members' having false beliefs. What *would* be surprising is that those beliefs should be ones on which the whole fabric of co-operation and communication depends.

Additionally, if we regard some of the arguments given earlier as successful in establishing nativism, then the innateness of folk-psychology provides us with good reason for believing it to be correct. For why should the beliefs in question be innate *unless* they are true? It might be replied that many innate features of organisms involve misrepresentations of the environment. For example, ducklings seem to have an innate belief that the first moving thing they see is their mother. Since this will lead them mistakenly to attach themselves to a human being, if a human is the first moving thing they see, it might be argued that there is no general reason for thinking that innate beliefs will be true. But one difference between the two cases lies in the relative stability of the duckling environment. (Another is that the mechanism in the duckling is probably not properly characterized as a *belief*. Are we really prepared to apply the practical-reasoning-model here?) Since the mother duck is almost always the first thing a duckling will see, it is easy to understand why the mechanism of imprinting should have been selected in evolution.

Human folk-psychology, in contrast, has to serve in a very wide and unpredictable range of circumstances. Much of the success of human beings as a species lies in their ability to adapt to almost any environment, as well as in their ability to accumulate knowledge about the world around them and hence adapt it, in turn, to human needs. So the innate basis of these abilities (namely our mutual beliefs about human psychology) has to be workable whatever the circumstances, and whatever other beliefs and practices are current in the community. It is hard to see how anything could serve this purpose *except* the truth.

I conclude that we have good reason to think that the generalizations of folk-psychology are both innately believed and true. Then, since it appears that the evolutionary selection of beliefs and belief-acquisition mechanisms (at least in creatures as complex and adaptable as ourselves) is a reliable process, folk-psychology will count as innately known. So we do have substantive innate knowledge of an aspect of the world, namely the psychology of members of our own species.

Folk-Psychological Concepts

What of the constituent concepts of our common-sense beliefs about the mind? Are they, too, innate? Or are they rather acquired from experience prior to the triggering of our innate beliefs? (Recall from chapter 4 that our possession of innate knowledge need not entail that we also possess innate concepts.) One account of mental concepts that might allow them to be derived from experience would be the Cartesian model. This holds that all mental states are transparently available to consciousness, our concepts of them being simple recognitional capacities. Then perhaps introspection of our own mental states might be sufficient for us to acquire our conceptions of them, even if it is not sufficient for us to learn of the causal relations between them, as we noted above.

The Cartesian conception of mental concepts is now almost universally rejected, for at least two reasons. The first is that it makes it very difficult to explain how we can have knowledge of the mental states of other people. If our conception of each type of mental state is given purely by its inner subjective quality, which we can of necessity know only from our own case, then it is problematic, to say the least, how we can know that other people have similar subjective feelings in similar circumstances.[9]

The second objection to the Cartesian account is that there are many types of mental state for which it does not even begin to look plausible. Consciousness of my own beliefs and desires, in

[9] For further discussion, see my *Introducing Persons*, ch. 1.

particular, surely cannot be a matter of immediate recognition. For *in virtue of what* would they be recognized? These states do not have distinctive feels to them, in the way that pains and other sensations arguably do. Indeed, since there are potentially infinitely many distinct beliefs and desires, of which I have concepts *before* I come to have them, there is a real problem here as to how I might be supposed to have acquired these concepts from introspection.

There is no unanimity as to what account of our mental concepts should replace the Cartesian one, but functionalism is now the most popular.[10] This holds that mental states are individuated by their *normal causal role*, or function. Indeed, the functionalist view is that mental concepts gain their sense from their position within the generalizations of folk-psychology. Knowing what a belief is, for example, is a matter of knowing how it tends to interact with other mental states such as desires and decisions.

There are problems with functionalism that need not detain us here.[11] Suffice it to say that if functionalism is correct, then mental concepts, in being acquired along with folk-psychological beliefs, will be innate if the latter are. Indeed, it looks as if the same will be true on any viable alternative to Cartesianism. For, as Chomsky points out, our mental concepts appear to be connected to one another in very many complex and subtle ways.[12] This means that it will be hard to explain how these concepts are acquired so fast, and on such slender exposure, unless they are supposed to be innate.

There is one issue that we do need to pursue further, since it bears on the likely correctness of folk-psychology. This is the question whether, in gaining their sense from being embedded within folk-psychological beliefs, mental concepts are employed as *natural-kind terms*. (A natural-kind term is one that is used with the intention of 'dividing nature at the joints', in Plato's phrase—that is, of picking out a class of things that would be recognized as such in a completed science.) To take the view that mental concepts are employed as natural-kind terms is to

[10] See the papers collected in pt. 3 of Ned Block (ed.), *Readings in Philosophy of Psychology I* (London: Methuen, 1980).
[11] For further discussion, see my *Introducing Persons*, chs. 4 and 6, as well as many of the papers mentioned in n. 10 above.
[12] *Language and Problems of Knowledge*, 26–34.

think of folk-psychology as a primitive form of scientific theory of the mind, as its very name suggests. Then if, as is possible, a completed scientific psychology would not employ such notions as 'belief', 'desire', or 'pain', it will turn out that these terms fail to refer (just as 'phlogiston' fails to refer), and that the whole of folk-psychology should be rejected as false.

However, whether or not a body of beliefs constitutes a scientific theory is at least partly a matter of the intentions with which it is held. In particular, we have to be prepared to abandon the theory if a better one emerges. But we do not have this attitude towards our common-sense psychology. If a future scientific psychology finds that it can operate best without such notions as 'pain' or 'belief', we will not accept that there are no such things as pains and beliefs, any more than we accept that tables are not really solid objects in the light of the discoveries of modern physics. Rather, we will continue to operate with our folk-psychological notions and generalizations *alongside* scient-ific psychology. While I have argued that folk-psychology is a largely correct theory of the mind, it does not follow that it is a scientific theory of the mind. (Compare folk-physics, which includes such truths as 'The chair broke because David was too heavy for it'. We retain this as true, despite the fact that 'chair' does not designate a natural kind.)

My conclusion is that we do indeed possess a body of reliably caused true beliefs that are innate, namely our beliefs about our own psychology.

9

Powers of Mind: The Core of Empiricism

In this chapter I shall consider whether the essential concerns of empiricists can be characterised so as to be consistent with nativism. If they cannot, then empiricism should now be pronounced dead, given the strength of the nativist case.

The Problem of Unity

Which of the two main strands in empiricism (opposition to nativism, and opposition to the substantive a priori) constitutes its true core? Or should the essence of the empiricist project be described in some third way that would embrace both? Clearly the two strands are at least partly independent of one another. One could believe in the existence of substantive a priori knowledge without endorsing nativism. One could, for example, believe in a special faculty of intellectual intuition, as many platonists have done. This would accord us knowledge of the abstract realm in a quasi-perceptual manner. Although the faculty itself would be innate (just as the faculty of vision is innate, even for an empiricist), the beliefs that it gives rise to would not be. Then, since belief in the substantive a priori does not entail a belief in nativism, the sort of attack that empiricists have mounted on the substantive a priori cannot merely have been part of their campaign against nativism.

In contrast, knowledge that is innate would at the same time be a priori, at least in the sense of not being learned from experience. So to accept that there is substantive innate knowledge would commit one to belief in some forms of substantive a priori knowledge. It might then be suggested that the empiricist opposition to nativism may be motivated by their opposition to the substantive a priori. However, this would hardly warrant an attack on nativism as such, of the kind that we find in Locke. For an attack on the substantive a priori would not, on any account of the matter, be an attack on all forms of innateness. It would only be innate knowledge that would be undermined, leaving innate beliefs, concepts, and information-bearing mental structures untouched. Yet empiricists have characteristically denied all these ideas also. So their motivation cannot have been a mere consequence of their opposition to the substantive a priori.

It might be suggested that I have erred in characterizing empiricism purely negatively, as involving both the denial of nativism and the denial of substantive a priori knowledge. Perhaps a positive characterization might embrace both. Indeed so: that is just what I am now seeking. But we cannot take the obvious (and traditional) option, and claim that empiricism consists in the demand that all substantive knowledge must be grounded in experience. While such a demand does indeed rule out both innate knowledge and a priori knowledge, we should still be unable to explain the empiricist opposition to innate concepts, beliefs, and information-bearing mental structures, none of which need count as knowledge. Since concepts, for example, might be innate although no knowledge is, the empiricist insistence that all knowledge must be grounded in experience cannot explain their belief that all concepts, too, must be derived from experience.

Nor will it help to suppose that the basis of empiricism consists in the positive demand that all concepts should be acquired from experience. While this may be thought (wrongly, as we saw in chapter 4) to entail that there can be no innate beliefs or knowledge, it cannot explain the empiricist opposition to information-bearing mental structures. Nor is it obvious why it should be thought to rule out the possibility of substantive a priori knowledge. For it might be maintained that we can arrive at substantive knowledge of some aspect of the world by thought alone, despite those thoughts employing empirically acquired concepts. For example, someone could claim that while mathematical *concepts* are abstractable from experience, mathematical *knowledge* depends rather upon a special faculty of intellectual intuition.

Moreover, both of the above proposals would still leave us in need of an account of the motivation behind the positive demand. Why *should* one believe that all knowledge, or all concepts, must be grounded in experience? Indeed, for all that has so far been said, it may turn out on investigation that the positive characterizations given above serve merely to mask an underlying disunity of motive. In which case we should have made no progress with our question whether empiricists, while remaining true to their essential concerns, can be brought to an acceptance of nativism.

A rather different proposal might be that it is phenomenalism that is the core empiricist commitment. For if all knowledge and thought must in the end reduce to patterns within our own subjective experience, then presumably there can be no substantive a priori knowledge, nor any knowledge that is innate. Moreover, if the experience in question consists of unconceptualized and unstructured sense-data, as phenomenalists have traditionally maintained, then presumably neither concepts nor information-bearing mental structures can be innate either. But this suggestion has already been discussed, and rejected, in chapter 1. Rather, the fact that many empiricists have embraced phenomenalism is best explained in terms of their commitment to the two main negative strands already mentioned, together with their foundationalism.

A final suggestion might be that the basic concern of the early empiricists was with matters of justification. Perhaps all that unifies opposition to nativism and to the substantive a priori is that neither belief in innate concepts or knowledge, nor the belief that we may obtain substantive information about the world through reason alone, is sufficiently justified to be acceptable. In one way this proposal converges with my own hypothesis, to be developed shortly. But I do not see how it can serve, on its own, to capture the nature of the empiricist objections to nativism and the substantive a priori. For we should still want to understand just *why* all claims to such knowledge must be insufficiently justified.

A Historical Hypothesis

My own suggestion is quite different. I propose that the core of empiricism should be seen as lying in the idea that epistemology is constrained by science, and by psychology in particular. In my view, the most basic empiricist commitment is to the thesis that claims to knowledge should only be granted on condition that they can be rendered consistent with our best theory of the powers of the human mind, and of the mind's natural modes of access to reality. No knowledge-claims are to be allowed, except where we can provide at least the beginnings of a naturalistic account of the processes through which that knowledge is acquired. (A *natural* process is one that falls under causal laws. It is one that can in principle be subsumed within the laws of nature, whether those laws are known by us or not, and whatever form they might ultimately take; it is not presupposed that all natural processes are physical, for example.) But this is not to say that epistemology then becomes absorbed into natural science.[1] Rather, it is to place an additional normative constraint on epistemology—namely, that we should be able to see how our claims to knowledge might be fitted into the framework of a natural science.

On this account, the basic reason why early empiricists denied the existence of innate knowledge and concepts was that the only theory available to them at the time, of the process through which an item might come to be innate, was a *non-natural* one, namely direct intervention by God. Although the early empiricists would have granted that intervention by God was conceptually possible, it conflicted with their overall attempt to fit our idea of ourselves and our knowledge into a broadly scientific framework.

If this were really the reason why the early empiricists rejected nativism, then why did they not say so? It may be that what I am calling their core commitment formed such a fundamental part of their outlook as to be almost invisible. Alternatively, their real argument may not have been easily expressible in public, for political reasons. Most early empiricists were also theists, and even those who were not had to respect the immense political power of the Church.[2] An attack upon nativism on the grounds that it requires us to believe in God's

---

[1] This is Quine's position. See *From a Logical Point of View*, and also *Ontological Relativity* (New York: Columbia University Press, 1969).

[2] Witness the lengths to which Hume had to go in disguising his real views in *The Dialogues Concerning Natural Religion*.

intervention in the human mind might have seemed like a direct attack upon theism. For if God exists, why should he not intervene in the natural world if he chooses to do so? There is really no *argument* for insisting that natural events admit of naturalistic explanations, except the success of science in the long run. I suggest that instead of facing this issue head-on, and openly declaring their commitment to the explanatory adequacy of science, the early empiricists may have chosen either to argue against nativism on quite other grounds, or to assume its falsity and render the hypothesis of divine intervention unnecessary, by providing an alternative account of the genesis of human knowledge through experience.

Note that my proposal is consistent with reliabilism as a theory of knowledge. For the thesis in question belongs to epistemology, being concerned with what we may reasonably believe ourselves to know (that is, with second-order knowledge), rather than with first-order knowledge as such. An empiricist can grant that there may, as a matter of fact, be knowledge of which we can provide no naturalistic account. For we may have true beliefs that are in fact caused by some reliable process, but where the process in question is unthinkable in terms of current science. In such a case we would indeed have first-order knowledge, but an empiricist would insist that we nevertheless have no right to rely upon the beliefs in question. Those beliefs should at least be suspended (we should hold back from the second-order belief that they constitute knowledge) until we can begin to provide some naturalistic account of their mode of acquisition.

Explanatory Advantages

The main advantage of my proposal is that it enables us to unify the early empiricist objections to nativism and to the substantive a priori (particularly platonism). In both cases knowledge-claims were rejected because there was available no naturalistic account of the belief-acquisition processes involved.[3] In the case

---

[3] This is why the belief that we have either innate or substantive a priori knowledge turns out to be unjustified, thus converging with the justificationalist account of empiricist motivations outlined earlier. For in the absence of any

of nativism, the hypotheses that either knowledge, concepts, or information-bearing mental structures are innate were rejected because they seemed to require non-natural intervention in the human mind by a veracious God. In the case of the substantive a priori, the objection was that there could be no naturalistic explanation of how reason could have acquired the power to generate knowledge of anything outside itself. It would apparently have required God's intervention to ensure that the structure of our reason accurately mapped the structure of the appropriate mind-independent realm. The platonist hypothesis of a special faculty of intellectual intuition, for example, was rejected because we cannot even begin to give an account of the psychological structure of such a faculty, or of how there could be causal contact between necessarily existing abstract entities and the human mind. Yet the only other alternative open to a platonist, namely the suggestion that our knowledge of the abstract realm is innate, brings us back to a non-natural model of belief acquisition once again. God's intervention would apparently have been needed to ensure that our innate beliefs were true of the abstract realm.

My proposal also coheres well with the fact that the early empiricists immersed themselves in enquiries that were quite explicitly psychological. It is because they thought that the theory of knowledge had to be fitted into a broadly scientific outlook, and rendered consistent with our best scientific theories, that they regarded epistemological and psychological questions as belonging essentially together.[4] Many commentators, in contrast, have regarded the close conjunction of psychological with epistemological enquiry as being seriously confused. They have said that it is one thing to ask how the mind actually works, and how beliefs are actually formed, and

naturalistic account of our mode of acquiring such knowledge we are not justified in believing it to exist, given a general commitment to the explanatory adequacy of science—see the discussion in the final section of this chapter.

[4] Their motivation is therefore slightly different from that of Goldman in *Epistemology and Cognition*, whose approach is otherwise very similar. Goldman thinks psychology is relevant to epistemology because, as a reliabilist, he thinks it necessary to investigate how reliable our belief-acquisition processes actually are. The concerns of the early empiricists were broader, in that they were attempting to fit our picture of ourselves and our knowledge into a scientific outlook.

quite another to ask what we can know, and how. The one is a factual, the other should be a normative, enquiry. On my reading of empiricism, however, these questions are intimately related to one another. For if epistemology is constrained by psychology, then we cannot begin to settle the question of what we can know without at least sketching the outline of the various faculties and psychological processes involved in the acquisition of our beliefs.

My account also fits neatly with the ways in which both Locke and Hume explain the motivation of their work, as I shall show in a moment. Berkeley, however, might seem to represent something of a problem for my reading of empiricism, given the central place accorded to God in his philosophy—though even here the role of God is limited to supplying the *data* of experience; God is not supposed to intervene directly in the human mind. But, in fact, Berkeley's motives were indeed different. He adopted empiricist premisses in order to serve his own theological purposes, rather than as part of an attempt to fit our view of ourselves and our knowledge into a broadly scientific framework. Instead, his project was to defend theism from what he took to be the twin threats of materialism and atheism. He therefore ought not to be counted as an empiricist at all, on my account.

Now consider the case of Locke. In the Epistle to the Reader of the *Essay* he tells us how, in the course of discussions with friends on an unspecified topic, they came to feel that if they were to make progress with it they should first examine their own mental powers. The task that Locke then set himself was to see what subjects the human understanding was or was not fitted to deal with. In the Introduction he then writes in similar spirit, thus:

This was that which gave the first rise to this *Essay* concerning the *understanding*. For I thought that the first step towards satisfying several inquiries the mind of man was very apt to run into was to take a survey of our own understandings, examine our own powers, and see to what things they were adapted.

The project, as I see it, was to attempt to settle disputes concerning the extent of human knowledge by first providing an outline of the faculties of the human mind, and of the mind's natural modes of access to the world, and then to regard knowledge-claims as constrained to be consistent with that account. This is precisely what I am maintaining is the core empiricist commitment.

(A problem for this reading of Locke's project is that it has seemed to many commentators that the naturalism, and rejection of nativism, of the *Essay* are inconsistent with the doctrine of Natural Law that lies at the centre of Locke's *Two Treatises of Government*. For example, at one point in the latter he speaks of Natural Law as being 'written in the hearts of all mankind',[5] and certainly God plays a crucial role in Locke's moral system as a whole. Now one response to this would be to concede that the two works are indeed inconsistent, perhaps appealing to Locke's inchoate perception that they are to explain why he was so insistent that the *Two Treatises* be published anonymously. But, in fact, Locke's considered view was not that the principles of Natural Law are innate, but rather that they may be known by all mankind through the natural use of reason. Similarly, he thought that we may arrive at knowledge of God by means of rational reflection on observed facts about the world. Indeed, Locke's project throughout was to show that human knowledge—whether of science, or of theology, or of morals—is a natural phenomenon, arising from the application of our reason to the data given to us in experience.[6])

Similarly Hume, in describing the main aim of his work in the Introduction to the *Treatise*, argues that there is a sense in which the science of human nature lies at the foundation of all the sciences, and that we may hope to make progess with the latter by studying the former first. Now he can hardly have meant (can he?) that we should expect particular discoveries in physics or chemistry to be consequent on advances in psychology. Rather, he is more plausibly read as saying that claims to knowledge, in general, have to be rendered consistent with the powers of mind ascribed to us by our best psychological theories—which is again the core empiricist commitment. Even

---

[5] See *Two Treatises*, ii, sect. 11.

[6] This is shown convincingly by Stephen Buckle, *Natural Law and the Theory of Property* (Oxford: Oxford University Press, 1991), ch. 3, who draws heavily on Locke's *Essays on the Law of Nature*.

more explicitly, in the opening section of the *Enquiry Concerning Human Understanding* Hume writes:

The only method of freeing learning, at once, from these abstruse questions [of metaphysics], is to enquire seriously into the nature of human understanding, and show, from an exact analysis of its powers and capacity, that it is by no means fitted for such remote and abstruse subjects.

Here again the project is the core empiricist one, to constrain knowledge-claims by our best theories of the mind's natural modes of access to reality.[7]

Notice that on the account being proposed here, the traditional empiricist insistence that all knowledge must be grounded in experience turns out not to belong to the foundations of empiricism as such. Rather, the early empiricists stressed the role of experience in the acquisition of knowledge because experience was the only belief-acquisition process of which they could begin to give a naturalistic account. So my proposal not only unifies the empiricist opposition to nativism and to the substantive a priori, but also sees these as flowing from the same underlying commitment as does the empiricist lauding of experience.

Should we be Empiricists?

Let us now ask whether the principle of charity, as well as textual fidelity (broadly construed), favours my interpretation of the basis of the empiricist tradition. Put differently, this is to ask whether we have any powerful reasons for thinking that knowledge-claims should only be endorsed where we can provide some sort of naturalistic account of the relevant belief-acquisition process. Which is as much as to say: should we, today, be empiricists (given my proposed account of the core of

---

[7] Note how the above quotations from Locke and Hume seem to echo the Critical Philosophy of Kant. Yet the way in which Kant takes up the empiricist challenge—which is to explain how reason can have the power to generate substantive knowledge—is not a naturalistic one. It is rather to claim that the basic structure of the world, as object of our experience, is determined by the subjective constitution of the human mind. He also differs in thinking that we can have access to this constitution a priori. For some further discussion, see the final section of chapter 10.

empiricism)? Part of the point of asking this question is that it will enable us to approach the issue of what *motivates* what I am calling the core empiricist commitment. Why would anyone want to constrain knowledge-claims by our best psychological theories?

One line of thought is that not only is my construal of empiricism consistent with a reliabilist theory of knowledge, as I argued above, but that it may actually be entailed by such a theory. For when we do epistemology we want to know what we can know. This means, for the reliabilist, that we want to know which beliefs or belief-types are caused by reliable processes. But a process that leads us to form beliefs inconsistent with our other beliefs cannot in general be reliable, if we suppose that our beliefs are on the whole true. So a belief in the reliable acquisition of one of our first-order beliefs will not itself be reliably acquired (nor constitute second-order knowledge) if it conflicts with our other beliefs, including those of science. This then suggests that we can only know that we have knowledge of something, if we can provide at least the beginnings of a scientific account of *how* we know that thing— which is precisely empiricism's basic demand.

It might be objected that it is one thing to say that my belief in the reliability of a given belief-acquisition process is not in conflict with my scientific beliefs, and quite another to say that I must therefore be able to provide a scientific account of that process. For surely a belief can be consistent with a body of scientific theory without being explicable in terms of it. The absence of conflict might plausibly be thought to be required by reliabilism. But it is the demand for scientific explicability that constitutes the core of empiricism, on my account, and that we are now attempting to justify.

The distance between these two requirements may, however, be smaller than it appears. For recall that the empiricist is insisting only on the *beginnings* of a naturalistic account of the relevant belief-acquisition process. It may then be that in many cases, at least, the reason why we cannot provide even the beginnings of such an account is that current scientific theory strongly suggests that there is *no such process*. So while reliabilism may not in itself entail the empiricist constraint on knowledge-claims, it will do so when put together with the thesis that current science does not recognize the existence of the belief-acquisition process in question.

But can it really be reasonable to deny that we have knowledge of some subject-matter, merely because our present scientific beliefs provide us with no materials with which to frame a remotely plausible account of how we might have acquired that knowledge? Let us consider an example in some detail, as a test case. Suppose that some person, or group of persons, claims to be *prescient*. That is to say, they claim to have intuitive (non-inferential) knowledge of the future. Of course many people have actually made such claims, but let us imagine an example in which it seems indisputable that the predictions very often turn out right. So these people apparently have true beliefs about the future that are not arrived at by inference from current tendencies. The question, then, is whether we can give some account of the process by which those beliefs are acquired; and, if not, whether we are justified in denying that these people may be said to have knowledge of the future.

I can think of at least three hypotheses that an empiricist might propose, in order to account for all those cases of the apparent phenomenon of prescience that have actually occurred. First, it might be suggested that the people in question are predicting the future on the basis of a *non-conscious* inference from their knowledge of current tendencies, in which case there would be nothing mysterious about their powers. But this may turn out not to be the case. For we can easily imagine that they may be able to predict future events that could not possibly have been deduced from current tendencies, such as the accidental death of a particular person in a car crash in two years' time, perhaps also predicting the date and place of the event.

Secondly, there is the possibility of fraud. It may be suggested that those who claim prescience are covertly bringing about, themselves, the events that they predict. But again this might conceivably be ruled out, either because the events in question are beyond the causal powers of any individual or group of individuals (such as the appearance of a new comet in the sky on a particular date), or because we take steps to ensure that those who make the predictions have no opportunity to interfere with the course of events, perhaps by imprisoning them.

It might finally be suggested that the predictions are framed with sufficient vagueness to give merely the *illusion* of accuracy. This idea is already familiar from astrology, which purports to predict features of people's character and life history from the distribution of the planets at the time of their birth. For provided that the predictions are sufficiently vague and general, and concern topics that people want very much to hear about, and that occur fairly commonly, there is a good chance that those predictions will come to be regarded as having been fulfilled. (If your horoscope for the week says 'Personal relationships run smoothly', you may, if you are inclined, regard it as verified by the fact that you get on well with your lover all week, ignoring the fact that if it had said the opposite you would equally have regarded that as true, in the light of the row you had with your mother on the Wednesday.) An empiricist could suggest that something similar may also be taking place in our imagined case of prescience. But again it may turn out not to be so. For the predictions in question may be quite precise, concerning matters that are perhaps unusual, and of no particular human interest. It would then be hard to see how our impression of success could be illusory.

If a situation of this sort were to occur, it would be a serious embarrassment to an empiricist, given my characterization of their position. For we have not the faintest idea how there can be a reliable process for acquiring true beliefs about the future, except by inference from current tendencies. For how could a future event exert any kind of influence on the human mind? How could the mere fact that an event *will* take place at a particular time in the future give rise to any process bringing someone to believe that it will? The idea seems barely intelligible. Yet, in the example above, we would have powerful reasons for thinking that these people do have knowledge of the future, based upon their past success. So, in such a case, we could know that they have knowledge, having every reason to think that their beliefs about the future are in fact produced through a process that is reliable, although we cannot even begin to give an account of the nature of that process.

Now in one respect this example provides a tougher opponent than empiricism has traditionally faced. This is because there is no obvious way of reinterpreting the content of the beliefs in question so as to render their mode of acquisition less problematic. In connection with logic and mathematics, in contrast, one standard empiricist move has been to deny that we are forced to interpret the propositions in question as being concerned with a mind-independent realm of abstract objects. If we can rather construe such propositions as being concerned with mental constructions of one sort or another, then there will be no special problem in explaining how there can be natural processes that lead us to have knowledge of their truth. In the case of propositions about the future, on the other hand, no such reinterpretation is available.

We could therefore put forward a weaker version of the proposed core of empiricism, which might enable us to distinguish between prescience on the one hand, and the various forms of platonism on the other. We could say that no claims to knowledge should be granted where we cannot begin to give a natural account of the process through which we acquire that knowledge, *unless* the evidence for the existence of some sort of reliable process is overwhelming, and *unless* there is no possibility of reinterpreting the content of the beliefs in question in such a way as to render their mode of acquisition less problematic. So in the end (if sufficient evidence of the sort described above were to emerge) we may have no choice but to accept that we can have knowledge of the future through prescience, even though we cannot begin to account for the mechanism involved.

The Adequacy of Science

In fact we may also respond to the example of prescience more directly. We can insist that it is merely imaginary, and that a genuine case of prescience will never really occur. For the empiricist need not be claiming to know a priori that all knowledge must arise through natural processes, of which we can in principle provide an account. Rather, their attitude should be seen as resulting from a more general commitment to the ultimate (or at least in principle) explicability of all natural phenomena, including that of belief acquisition. This commitment may be supposed to receive its justification from past scientific success. An empiricist may therefore respond to our example by saying that they are prepared to bet that a genuine case of prescience will never in fact occur—precisely because we cannot begin to see how there could be any natural process underlying the acquisition of our beliefs in such a case.

It may be objected that empiricists, as such, cannot be committed to the thesis that all processes in nature are in principle explicable by science, happening in accordance with causal law. For Locke, for one, believed that there are natural processes—particularly those connecting physical events in our bodies with ideas in our minds—that must forever outstrip our powers of explanation.[8] But there are two points to be made in reply. The first is that Locke was unduly sceptical about our abilities to discover the hidden processes at work in nature, as subsequent scientific advance has shown. The second, and more important, is that he was working with a much more demanding concept of explanation than that presupposed here. In Locke's view, connections between events will only count as having been *explained* if they have been made fully and rationally intelligible to us. He might then have been happy to agree that all processes in nature occur in accordance with causal laws that are in principle discoverable, even though appeal to these laws can never make events intelligible in the way that 'Anything red is coloured' is intelligible.

Looked at in the way I am suggesting, empiricist constraints on knowledge-claims may be seen to stem from a more general belief in the ultimate adequacy of science. The sequence of thought would be this: if we were to possess knowledge of the matter in question (the future, say), then our beliefs would have to be caused by a reliable process; but if our current science is such that we cannot even begin to frame a hypothesis as to what that process might be, then this in itself provides us with good reason for doubting its existence.

An analogy may help here. Suppose someone suggests that zebras in the wild put on overcoats at night to keep warm.[9] Are we not prepared to bet, in the light of our current knowledge, that this will never turn out to be true? Indeed, if someone were

[8] See the *Essay*, bk. IV, ch. III.
[9] I have lifted this example from Daniel Dennett's *Brainstorms* (Brighton: Harvester, 1978), though he uses it for quite another purpose.

to present evidence of its truth, should we not do our best to dismiss or explain that evidence away? For if it *were* true, it would apparently be wholly inexplicable. Are we to imagine that zebras have their own secret technology, which enables them to weave cloth? Or that they have an elaborate and so far undiscovered system for stealing overcoats from human beings? These ideas pass beyond the possibility of belief, given what we already know about zebras and their habitat. Similarly, I suggest, with our imagined case of prescience: given what we already know about the world, we may be sure that it will never happen.

This is not to say that the empiricist constraint on knowledge-claims is an infallible one. Plainly it cannot be, since it led the early empiricists to deny the truth of nativism—as it turns out, incorrectly. It may nevertheless be a reasonable one. If we share the empiricist belief in the ultimate explanatory adequacy of science (or at least their belief that all processes in nature are natural ones, happening in accordance with causal law), then we shall deny that there are any natural phenomena for which there is no natural explanation. We therefore have reason to deny that some suggested phenomenon will ever in fact occur, if to the best of our belief there can be no natural explanation of it. Into this category, in my view, fall not only claims of prescience, but also astrology, magic, and various alleged psychic phenomena.

We are now in a position to articulate the underlying motivation behind the core of empiricism. It consists, first, in the search for an explanatory coherence within the overall system of our beliefs; and, secondly, in a commitment to the explanatory adequacy of science. The first of these strands leads empiricists to try to make our first-order beliefs about the world cohere with our best theory of the human mind and its powers, where necessary (and if possible) reinterpreting the subject-matters of those beliefs to render them more accessible to the human mind. The second strand leads them to reject claims to knowledge where we cannot even begin to construct a plausible naturalistic explanation of the manner in which that knowledge might have been acquired. Both strands are, now, eminently reasonable. It is obvious that we should try to weld our beliefs into a coherent system if we can (more on this in chapter 12).

And the early empiricists' methodological commitment to the explanatory adequacy of science has been amply vindicated by subsequent scientific progress.

However, that *we* are justified in believing in the explanatory adequacy of science, given the huge success of science, does not mean that the same can be said of the early empiricists. On the contrary. While their commitment to science was not a matter of blind faith, since notable advances in knowledge had already been achieved, it was not fully justified either. So here may be a further reason why the early empiricists did not make clear the true nature of their opposition to nativism. For, had they done so, they would have had to admit that their position was almost as reliant upon faith as that of their theistic opponents.

It appears that my proposal concerning the core of empiricism can indeed be supported by considerations of charity. If we cannot even begin to give an account of the process through which we might have acquired a set of beliefs, let alone an account that shows the process to be a reliable one, then it is surely only reasonable that we should refrain from committing ourselves to those beliefs if we can possibly do so. And if we cannot avoid the conclusion that the beliefs in question are true ones, then we should, if we can, provide an interpretation of their content that would render their mode of acquisition less problematic. This is, I suggest, precisely what the empiricist maintains.

Now, if I am correct in my characterisation of the core of empiricism, it follows that contemporary empiricists should have no objections to evolutionary versions of nativism. For, unlike divine intervention, the selection of innate characteristics through evolution is certainly a natural process. Indeed, it is one of which we not only have the outline of an account, but a well-developed scientific theory. Moreover, evolutionary selection is very probably a reliable process where the determining of innate belief is concerned, if the points made in chapter 8, concerning the survival value of true belief, were sound ones. As empiricists, we therefore have no principled reason for denying the existence of innate knowledge. Since we can indeed provide an account of the process through which a belief might come to be innate, the empiricist constraint on knowledge-claims has no application.

10

Evolutionary Nativism and A Priori Knowledge

In this chapter I shall consider how much remains of the traditional conflict between empiricism and rationalism, given my proposed account of the core of the former.

Nativism and the A Priori

In chapter 9 I argued that contemporary empiricists should have no objection to evolutionary versions of nativism. It follows that they should also have no objection to some forms of a priori knowledge. For knowledge which is innate will at the same time be a priori, at least in the sense of not having been learned from experience. This is, if correct, a startling result. It means, not only that one can remain an empiricist while accepting that there are innate information-bearing mental structures, innate concepts, and innate knowledge, but also while accepting the existence of substantive a priori knowledge. Indeed, you may be inclined to object that this is too good to be true. For there would appear to be a real danger, now, that I have construed empiricism in such a way that the differences between empiricism and rationalism may disappear altogether.

For example, if it were to turn out that an evolutionary nativism can provide an adequate defence of platonism, then on my account a contemporary empiricist would probably have no objection of principle to *any* of the traditional rationalist doctrines. In fact it would turn out that the early empiricists' opposition to rationalism derived entirely from the poorly developed state of science at the time, in particular from the absence of any sort of theory of evolutionary selection of inherited characteristics. Some might find such a conclusion so implausible as to warrant a rejection of my account of empiricism. Others, accepting that account, might take the conclusion to mark the demise of empiricism as a distinctive movement in philosophy. I shall make no attempt to decide which of these responses might be the more reasonable, since I believe that much does in fact remain of traditional disputes between empiricists and rationalists, even given my account of the essential concerns of the former.

If the course of evolution has provided us with substantive innate knowledge, either of our own psychology or of other aspects of the natural world, then it will at the same time have provided us with substantive a priori knowledge. It is important to realize, however, that the status of such knowledge as a priori will be quite different from what rationalist philosophers have generally had in mind when they have employed the phrase 'a priori knowledge'. In particular, the knowledge in question will not have been acquired merely through the thinking of it, by a process of reasoning alone. Innate a priori knowledge of folk-psychology, for example, would not consist of beliefs that can be discovered to be true just by thinking of their subject-matter; nor would their content be intuitively certain. Rather, we would find ourselves with those beliefs already; and then, provided that they are both true and reliably acquired through evolution, this would be sufficient to constitute them as knowledge. Hence at least one remaining difference from traditional rationalism is that the most plausible cases of innate knowledge for a contemporary empiricist will not be knowable by thought alone.

A related point is that the question whether a given belief is a priori, in the nativist sense, will not itself be an a priori one. It will rather be empirical. For recall our rejection of the KK thesis (that knowledge requires knowledge that you know) in chapter 5. Knowing something (whether a priori or not) is one thing, knowing that you know it may be something quite different. In the case of folk-psychology, for example, if our beliefs are, as a matter of fact, both innate and reliably acquired through evolution, then they will be known a priori (in the sense of not having been learned from experience). But to know their status as such will require empirical investigation, and argument from contingent premisses, such as we provided in chapter 8.[1] This is

---

[1] I assume—surely plausibly—that if someone innately knows something, they will not also innately believe that they know it.

in stark contrast with the traditional rationalist conception of a priori knowledge where, since the belief in question can be known to be true by thought alone, it is often felt that the knowledge that we know it a priori can also be established by the very same method.

With these remaining differences between rationalism and my contemporary empiricism established, let us turn to consider those domains of belief that have traditionally been regarded as a priori, such as mathematics and logic. And let us consider the platonist construal of the subject-matter of these disciplines, which entails that our knowledge of them is not only a priori but genuinely substantive, being concerned with a reality independent of our minds. Our question is going to be whether an evolutionary nativism can be pressed into the service of platonism, to provide a naturalistic account of how the process through which we acquire beliefs about the abstract realm may be a reliable one. But first let us consider what such a nativism might look like.

Perhaps the most plausible form of nativistic platonism is that defended by Jerrold Katz.[2] His approach combines evolutionary nativism with more traditional appeals to intuition as a warrant for our a priori knowledge of logic and mathematics. The advantage of this combination of views is that it is obviously not the case that we simply find ourselves with mathematical beliefs, as is perhaps so with folk-psychology. Any adequate account of mathematical knowledge should find some place for the concept of intuitively acceptable *proof* of new mathematical propositions. On the other hand, intuition, without the backing of a rich nativism, cannot be thought of as a form of direct access to the abstract realm, as we saw in chapter 3.

Katz's idea is that we have an innately given faculty of intuition, whose constituent structure mirrors that of the abstract realm. This faculty enables us to construct mental representations of items in the abstract world, and to intuit, on this basis, what has to be true about those things. But intuition is not to be thought of on the model of sense-perception. His claim is not that abstract objects exert a causal influence on our minds via the faculty of intuition, those objects being directly responsible for our beliefs. It is rather that the innately given structure of the faculty of intuition causes us to have true beliefs

[2] See his *Language and Other Abstract Objects*, ch. 6.

about items in the abstract realm, by determining our intuitive judgements when we manipulate our mental representations of those things. On this account our knowledge of the abstract realm is genuinely a priori. For it is arrived at by thought alone, this being the only process directly involved in our acquisition of new mathematical or logical beliefs. On the other hand, the intuitions we arrive at by such thought will only be reliable guides to the way things are in the abstract realm if the structure of the faculty of intuition has been reliably determined in evolution. So the knowledge that we know is not equally a priori.

What emerges from this discussion of Katz's platonism is this. Even if it were to turn out that empiricists can now accept the existence of substantive knowledge obtainable by reason alone, through accepting some form of evolutionary nativism, they can still maintain their traditional opposition to the characteristic rationalist thesis that we may know, by reason alone, that we are capable of obtaining substantive information about the world (or about ourselves) by reason alone. For they will continue to demand a naturalistic explanation of how reason can have acquired the power reliably to generate truths about things outside itself. It seems certain that no adequate answer to this can be provided, in turn, by reason alone. It will rather require the backing of some evolutionary (and hence empirical) explanation, if reason's claims to knowledge are to be allowed. So at least one further aspect of the traditional debates between empiricists and rationalists will survive my proposed account of the core of empiricism. Only if our a priori knowledge concerns nothing that is independent of our minds (that is to say, only if it is analytic) can we know a priori that we possess such knowledge.

Nativism and Mathematics

The only issue now remaining is whether an evolutionary nativism can be used to render acceptable to contemporary empiricists some of the particular beliefs traditionally defended by rationalists, such as platonism. For example, can a platonist respond to the challenge to provide a naturalistic account of the process through which we acquire knowledge of the abstract realm, by maintaining that such knowledge arises out of the structure of a faculty of intuition that has been determined through evolutionary selection? The difficulty for such accounts that I propose to raise first, relates to the supposed innateness of our knowledge of mathematics (the subject-matter of which is understood platonistically).

How can it be maintained that such knowledge is innate, given that a developed form of mathematics has only been in existence for a few thousand years? For it is obviously not the case that all humans now living are descended from a common ancestor, or single group of ancestors, of a few thousand years ago. On the contrary, human beings were already widely dispersed around the globe at the time when mathematical knowledge first began to make its appearance. Yet there appear to be no human beings who lack mathematical ability. Even those primitive tribes who (until recently) had no knowledge of mathematics at all (perhaps counting 'one, two, three, many') are able to grasp it quite easily when introduced to the ideas. So if mathematical knowledge is to be innate in its own right, it would have to be supposed that it evolved separately in all the different groups of humans at about the same time. This is extremely unlikely, given that evolution operates by random gene mutation.

One response to this problem would be to suggest that our innate knowledge of mathematics should really be subsumed within our knowledge of logic. For it is clear that all human communities, however remote in the past, will have had knowledge of the truths of logic (though whether those truths would concern platonic objects is of course another matter). Rational planning of a strategy of action would be impossible if one could not rely upon such truths as 'If I will not get A unless I do B, and I cannot do B unless I do C, then I will not get A unless I do C'. It would be tempting, then, for a platonist to maintain that it is the logic faculty that is most directly innate, the truths of mathematics being somehow derivable from those of logic. The cost of this response, however, is that it commits us to the logicist programme of reducing mathematics to logic, discussed in chapter 3. Yet few philosophers today believe that such a programme can succeed.

Chomsky, too, faces the difficulty of explaining how our knowledge of mathematics can be innate, since he is another who believes that it is so (but without endorsing platonism). His solution is to suggest that mathematics can be seen to be at least partially innate, by virtue of its similarities with the structures inherent in our innate language faculty.[3] In particular, the grammar of any human language will contain rules that may be applied recursively an indefinite number of times. (For example, the rule for 'and' enables you to reiterate conjunctions indefinitely: 'A and B and C and D and so on'.) Chomsky suggests that we may exploit this feature of natural language, which is innately known, in constructing a system of numbering.

Can a platonist, similarly, make use of Chomsky's suggestion? I believe not. For the idea is not that the truths of mathematics may be *reduced* to those of grammar. That would be absurd. Rather, it is that there is a formal *similarity* between the two, which can be exploited to enable us to construct the one on the basis of knowledge of the other. But a platonist, of course, cannot accept such an account as it stands, since platonists do not believe that mathematical truths are constructed. Rather, those truths are held to relate to an objective abstract realm, which exists independently of our minds. So a platonist would have to claim that the evolutionary selection of the language faculty is also reliable in giving us knowledge of mathematics, in virtue of the formal similarities between the structures of grammar and the structure of the mathematical realm. But in fact a mere similarity between a realm to which we do have reliable access, and one to which we do not, cannot be sufficient to confer knowledge of the latter.

For example, my perceptions are generally reliable in giving me beliefs about the features of the planet Earth. Now let us suppose that there is another planet in the galaxy that is in fact structurally similar to Earth. Do I thereby have knowledge of the features of that planet? Obviously not, at least until I learn on other grounds that the two planets are indeed similar. Equally, then, in the case of mathematics: the mere fact that there are structural similarities between the mathematical realm and the rules of grammar cannot be sufficient for my innate knowledge of the latter to give me knowledge of the former,

---

[3] See his *Language and Problems of Knowledge*, ch. 5.

until I know on other grounds that the two domains are indeed similar (and in what respects they are similar). But in order to know this, of course, I should first have to know something of the mathematical realm—which is precisely what we were trying to explain.

There are therefore real difficulties, for a platonist, in appealing to evolutionary versions of nativism to explain our knowledge of mathematics. For on the one hand it is implausible that such knowledge should be innate in its own right. On the other, attempts to explain such knowledge in terms of our knowledge of logic, or in terms of our knowledge of grammar (both of which are very probably innate), are beset with difficulties. It is, however, worth considering whether there are any further—more general—arguments against nativistic platonism. To this topic I now turn.

Evolutionary Platonism?

The main problem for a platonist, as I see it, is to explain why natural selection of a faculty of intuition should be a *reliable* process, leading us to have true beliefs about the abstract realm. For what difference could the way things are in such a realm make to our chances of survival in the physical world? If we are to appeal to evolution to explain our innate knowledge of the abstract realm, then it must be supposed that possessing truths about such a realm will make a difference to our survival. Those early humans who happened (as a result of random mutations) to have true beliefs about such a realm would have to have had a better chance of surviving than those who did not. But what difference can abstract objects make to our survival? You cannot eat an abstract object, nor poison yourself with one. Nor can ignorance of an abstract object lead you to be eaten by a tiger or drowned in a flood. So how can true beliefs about the abstract realm (or a faculty whose constituent structure enables us to obtain true beliefs about that realm) have been reliably selected in evolution?

In reply, it might be said that mathematical truths are extremely useful to those who possess them. For example, they can enable you to work out what would be a sufficient store of food to see your family through a winter. Ignorance of a truth such as 'If we need 6 kilos of rice a day, we shall need 900 kilos of rice to last through a winter of 150 days' may lead someone to die of starvation. Something similar will be true of logical and grammatical truths as well—the former underlying all planning and reasoning; knowledge of the latter being what makes communication possible.

Empiricists have traditionally provided a number of explanations of the usefulness of mathematical truths, ranging from the suggestion that they concern internal relations between our concepts to the idea that they consist of high-level empirical generalizations. But what is hard to see is how facts about platonic objects could enter into the explanation of the usefulness of mathematics and logic. If abstract objects cannot causally affect the natural world, then presumably there is at most some sort of structural correspondence between the two realms, in virtue of which pure mathematics can become applied. Knowledge of these structural features of the natural world itself would then suffice for whatever practical advantage is to be gained from mathematics.

Thus it surely cannot be ignorance of the properties of the abstract objects 6, 900, and 150, as such, which would lead me to die of starvation in the example above. Rather, it would be ignorance of the general structural features of the physical world in virtue of which it is true that 150 parcels of 6 kilos of rice makes 900 kilos. It would therefore be sufficient for me to survive that I should know some general truth from which it follows that, in order to have a parcel of rice for each day of the winter, each of which when placed on the scales will tip the needle to the '6 kilos' mark, I must collect a quantity of rice that will tip the needle to the '900 kilos' mark. This is not knowledge of abstract objects as such, but rather knowledge of the manner in which bodies of matter in the physical world may be collected together or divided up. (The platonist view, in contrast, must be that these properties of matter are structurally isomorphic with the properties of objects in the abstract realm, so that knowing the latter thereby provides us with knowledge of the former.)

It therefore remains problematic why evolution should be reliable in selecting beliefs about the abstract realm. For it is not truths about that realm as such that are an aid to survival, but rather truths about the structural features of the physical world in virtue of which the truths of mathematics can be applied to that world. These truths about the physical world could possibly have been reliably selected through evolution. But to go beyond this, to have true beliefs about the abstract realm itself, would confer no additional advantage in survival. In which case it seems that a platonist cannot explain in evolutionary terms how knowledge of the abstract realm can come to be innate. This will then leave an empiricist, who rejects claims to knowledge that cannot begin to be explained, with sufficient reason for denying that we can have knowledge of mathematics, if its subject-matter is construed platonistically.

Now the point here is not that evolution cannot secure more than is necessary to meet the needs of survival. Plainly it can, if random mutations are what fuel the evolutionary process. It is rather that evolutionary selection of beliefs is only a *reliable* process to the extent that it is the truth of the beliefs that confers advantage in survival. When it comes to beliefs whose content goes beyond those structural features of the natural world that may matter for survival, natural selection is just as likely to produce beliefs that are wrong as right. It will not matter for survival if innate mathematical beliefs are false with respect to the platonic realm, provided that they work successfully in the natural world.

Remember that what evolution would in the first instance deliver would be a faculty for manipulating representations of platonic objects, finding some but not others of these representations intuitively acceptable. But since the survival value of truths about platonic objects derives entirely from the supposed structural correspondence between the natural and platonic realms, evolution could only be reliable in respect of those structural features of the natural world. That is, it would lead us to find $2 + 2 = 4$ intuitively acceptable when, and only when, it is true that things may generally be divided up, and combined, and counted in such a manner.

In fact we can easily imagine worlds where $2 + 2 = 4$ would not be selected for, but rather, if anything, $2 + 2 = 5$. For example, a world where 2 apples and 2 pears put into a lunch-box produce a fifth piece of fruit, or a world where the very act of counting the union of the two sets produces an extra member, and so on. So how could we know that our innate belief that 2 + 2 = 4 is not of this sort? Maybe, in the platonic realm, 2 + 2 is really 5, but the structural facts about the natural world are such as to make '2 + 2 = 4' apply to it.

The point here needs stating with some care. For if evolution selects a faculty of mathematical intuition in virtue of the access that the latter gives us to structural facts about the natural world, and if there is *in fact* a structural correspondence between the natural and platonic worlds, then evolution will be reliable in delivering us truths about the platonic realm. Hence we would, in these circumstances, have innate first-order knowledge of platonic objects. And we cannot here object, against the thesis that evolution is reliable with respect to the platonic domain, that if the mathematical facts had been different, evolution would still have selected for us the very same mathematical beliefs. For we cannot intelligibly suppose that truths that are necessary (as are those of mathematics) can be other than they are.

It seems that an appeal to evolution can, after all, explain how innate knowledge of platonic objects is possible, given a reliabilist conception of knowledge. But what appeals to evolution *cannot* do is provide us with any reason for thinking that there *is* a structural correspondence between the platonic and natural realms. For we cannot, in this context, take our knowledge of 2 + 2 = 4 for granted. While this proposition may strike us as intuitively obvious, our question is precisely whether we have any reason to think that our intuitions are a good guide to the states of the platonic realm. Appeals to evolution therefore cannot give us the knowledge (second-order) that we have knowledge of platonic objects. On the contrary, it can only generate scepticism on the issue.

If we had no other option but to accept the platonist interpretation of the subject-matter of arithmetic, then the only proper conclusion would be that we have no reason to think that we have any knowledge of that subject-matter (namely, of platonic objects). But of course there are other options available, as we saw in chapter 3. Appeals to evolution therefore cannot render platonism acceptable to a contemporary empiricist. While evolution may possibly explain how innate knowledge of platonic objects is possible, it cannot provide us with any reason for thinking that such knowledge actually exists. While this is the case, it will remain necessary for an empiricist—who insists that claims to knowledge should only be allowed where we can begin to see in natural terms why we are reliable in making those claims—to continue to reject platonism.

Other Rationalist Beliefs

It turns out that the empiricist objection to platonism remains intact, even given my construal of the core of empiricism, and the consequent empiricist acceptance of evolutionary nativism. But what of other aspects of the traditional debate between empiricists and rationalists? What of empiricist objections to the claims of rationalists that we can have a priori knowledge of the existence of God, of the existence and immortality of the soul, or of the freedom of the will? On my account, the most fundamental of these objections is that no naturalistic explanation can be provided of the manner in which we might acquire such knowledge. Can this, too, survive the empiricist acceptance of nativism?

Here, as before, the main issue is whether the beliefs in question, or the faculty of thought that enables us to intuit the truth of those beliefs, could have been reliably selected through evolution. For the sake of simplicity, let us confine ourselves to consideration of only one of these forms of nativism. Let us consider the suggestion that we possess an innate faculty of mind that enables us to discover the truth of the beliefs in question by thought alone. Our conclusions will then readily generalize, to cover the possibility of innate belief as well.

Consider, first, the suggestion that we possess an innate capacity for acquiring knowledge of the existence of God. Can we explain how such a faculty of mind might have been reliably selected through evolution? Note that we cannot appeal to the social or psychological benefits of religious belief in this connection (supposing that there are any). For such benefits would be independent of the *truth* of the belief in question. Selection of a belief on grounds unrelated to truth is not a reliable method of acquiring true beliefs. So even if the faculty enabling us to arrive at belief in God were innate, and that belief were true, still our belief would not constitute knowledge if the explanation of its innateness were in terms of benefits unrelated to truth.

The only truth-related way in which belief in God could conceivably have value in survival would be if such a belief enabled one to secure, through prayer, God's positive intervention in the world. But this violates the empiricist constraint that the explanation of a belief-acquisition process should be a natural one, mentioning only processes that operate in accordance with causal laws. For God's intervention in the world, for example to save a life, would involve a violation of such laws. So I conclude that even an empiricist who accepts nativism will still have an objection of principle to the rationalist claim that we can have a priori knowledge of the existence of God.

Consider, now, the suggestion that we have an innately given faculty of reason that enables us to discover a priori the existence and immortality of the soul, or the freedom of the will. Once again it is difficult to see how a faculty with such powers could have been reliably selected through evolution. For it is hard to see what possible difference the possession of true beliefs on these matters would make to our survival. (Just as in the case of belief in God, it is not sufficient to show that the *beliefs* in question have survival value: perhaps belief in immortality can keep us from despair, for example, and perhaps belief in freedom of the will is a necessary condition of our having moral beliefs. It needs to be shown that it is the possession of a *true* belief that aids survival, since otherwise the belief-acquisition process will not be a reliable one.) It therefore follows that an empiricist will object to such knowledge-claims, on the grounds that no naturalistic account can be given of the manner in which we might acquire that knowledge.

Many philosophers have claimed to be able to demonstrate the existence of the soul, for example, believing that they could establish the conceptual impossibility of our *lacking* a soul. In fact such arguments generally turn out to be unsound.[4] But even supposing that some such argument were to succeed, empiricists would have a further line of defence. For suppose that it had been demonstrated that we cannot conceive other-

---

[4] However, see my *Introducing Persons*, chs. 2 and 3, for an argument which comes quite close to success.

wise than that we possess a soul. An empiricist will still demand to know how we come to have a faculty of reason that is capable of discovering a priori the fundamental nature of ourselves. How can thought alone have acquired such powers? Unless some evolutionary explanation can be discovered, an empiricist will wish to exploit the gap between our inability to conceive of something being otherwise (a fact about us and our concepts), and the fact of something being so (a fact about the world). Even if we cannot conceive of ourselves otherwise than possessing a soul, this does not mean that we know of the existence of the soul. On the contrary, this latter knowledge should be denied of us.

It may be worth stressing in conclusion to this section that the empiricist objection is not to the idea that reason, as such, has value in survival. That would be absurd. It is obvious that the human ability to reason is at least one of the distinctive features underlying the success of the species. The objection is only to the survival value of a faculty of reason that would contain within itself the power to generate true beliefs about the matters generally beloved by rationalists, such as the existence of God or the soul.

Transcendental Arguments

Finally, let us consider how empiricists should respond to various forms of transcendental argument, which attempt to show a priori what must be the case about ourselves or the world if experience is to be possible at all. Such arguments were developed most famously by Immanuel Kant in his *Critique of Pure Reason*, though he has had many imitators. His position is often thought to provide a kind of synthesis of the empiricist and rationalist traditions. For example, suppose it were success-fully argued that it is a condition for the possibility of our experience that all events should have causes. That is, suppose it were shown to be conceptually necessary that all events must have causes, if there is to be any experience at all. Is this something that my sort of empiricist could in principle accept? The answer will in fact depend very much on how the conclusion of such an argument is to be understood.

Suppose, first, that the conclusion purports to be a truth about a mind-independent reality (many contemporary philosophers have constructed transcendental arguments of this sort). Then in my view an empiricist should object to the possibility of our having a priori knowledge of it. For the fact (supposing that it were a fact) that we cannot conceive of the world otherwise than governed by universal causality is not enough to establish that all real events do have causes. Nor is it at all obvious how a faculty for obtaining such knowledge could have been reliably selected through evolution. No doubt a true belief in universal causality might be useful in survival. But given that only a handful of people have ever employed their supposed innate faculty in order to demonstrate the existence of universal causality, it is hard to see what advantage the possession of such a faculty can have conferred through evolution.

Kant himself saw, at least partly, the empiricist challenge to rationalism. (He regarded it as a challenge to show *a priori* how reason can have the power to generate knowledge of anything outside itself, rather than, as I propose, as a demand for a *natural* explanation of reason's reliability on such matters.) He put forward his doctrine of Transcendental Idealism in an attempt to overcome it. Kant's view is that the structure of the human mind imposes an order on the world of our experience, and what we may know a priori is limited to that world, considered as an object of possible experience by us. On this view, our proof that all events have causes would be guaranteed to be reliable, because it is a structure that reason itself imposes on the course of our experience.

There are, however, two different ways in which Kant's Transcendental Idealism can be understood, each finding some basis in his writings. The first regards it as a sort of dual-aspect realism. On this view, there is a realm of objects existing independently of our minds (hence 'realism'). But those objects can be thought of under two different aspects—first, as possible objects of our experience (as *Phenomena*), and, secondly, as they are in themselves, independently of our experience (as *Noumena*). All our knowledge, whether empirical or a priori, would be confined to the realm of *Phenomena*. And the truth that all events have causes would be imposed upon *Phenomena* by the structure of our reason. But then there is one item of knowledge claimed on this account that can*not* be explained as a structure imposed by our reason on objects as *Phenomena*—namely, the claim that there are objects independent of our minds (the claim that there are *Noumena*). So if this is how Kant is to be interpreted, he is still wide open to the empiricist challenge.

Alternatively, Kant's Transcendental Idealism may be understood as a form of sophisticated phenomenalism. On this view, it cannot be known that there are objects independent of our minds, and all a priori knowledge concerns only patterns that the mind imposes on the course of our experience itself. On this account, at least one of the empiricist objections of principle to the existence of transcendental arguments collapses.[5] For the a priori knowledge that all events have causes, while being informative (that is, new), would not be genuinely substantive (that is, mind-independent). The truth that all events have causes would in fact only concern the course of our experience itself. Admittedly, this truth would be, to a degree, substantive, since it is a contingent fact that there should be any experience at all. (Recall from chapter 1 that knowledge may be substantive by virtue *either* of being contingent *or* of concerning some reality independent of our thoughts.) But the substantive content of the claim that all events have causes would not reach any further than the bare, but substantive, claim that there is experience. If this is how Kant is to be interpreted, then in attempting to surmount the empiricist challenge he has, in reality, surrendered to it. For he has failed to show how substantive a priori knowledge is possible. He would only have been prevented from seeing this because he does not distinguish clearly between knowledge that is informative and knowledge that is substantive.[6]

In summary, I conclude that my proposed account of the core of empiricism leaves much of the traditional dispute between empiricists and rationalists intact. Both sides may now accept the existence of innate knowledge. But they remain divided over

---

[5] There remains the difficulty of explaining how we may know *a priori* that reason does impose an a priori structure on any possible experience.

[6] See his very confused explanations of the analytic—synthetic distinction, for example at *Critique of Pure Reason* A.6–7, B.10–11.

the general claim that it is possible for us to know that we have substantive knowledge through thought alone, as well as over many specific claims to a priori knowledge.

11

Our Knowledge of the External World

In this chapter I shall begin to consider the significance of my proposed account of the core of empiricism for the problem of scepticism.

Empiricist Epistemology

What would an empiricist epistemology now look like, if my proposals concerning the essentials of empiricism were accepted? One issue here is whether a contemporary empiricist, who may accept the existence of innate knowledge, and hence of one form of substantive a priori knowledge, will have anything of interest to contribute to contemporary debates in epistemology. At least a partial answer may be provided from amongst the ideas defended in chapter 10. For the strictures against platonism, and indeed against any claims to knowledge where we cannot give a naturalistic account of the reliability of the belief-acquisition processes involved, remain in force. The empiricist constraint on knowledge-claims may be expected to have a distinctive impact on epistemology. Many claims to knowledge may be disallowed because of it.

The main issue I wish to consider, however, is whether my contemporary empiricism is any better placed to combat scepticism, thus allowing that we can have knowledge of the physical world around us. Traditional empiricism has been closely allied to scepticism, or at least to phenomenalism. While Locke thought that we could have knowledge of a physical reality independent of our minds, his arguments have generally been considered weak. Most subsequent empiricists, from Berkeley and Hume through to Russell and Ayer, have doubted whether we could have knowledge of a physical world that is independent of our minds. Instead, they have embraced some form of phenomenalism, maintaining that our ordinary talk about 'physical objects' may be reduced to descriptions of immediate experience, or sense-data, and to descriptions of recurring patterns within that experience. It is worth asking, then, whether the traditional empiricist commitment to phenomenalism is in any way connected with opposition to nativism. It may also be worth looking again at the pressures that have tended to push empiricists towards scepticism.

However, we need to be clear at the outset about the different levels involved in our enquiry. For, given a reliabilist conception of knowledge, of the sort defended in chapter 5, we shall in fact have knowledge in any domain where we possess reliably caused true beliefs. So, provided that there is in fact a physical world, and that the processes that give rise to our beliefs about that world (either perception, or natural selection, or some combination of the two) are in fact generally reliable ones, then we do indeed have knowledge of the world. But this will be of little help to us in combating scepticism. For the knowledge in question would be entirely first-order, whereas the knowledge that we want is second-order. We want to know (second-order) whether we do indeed have knowledge (first-order) of the physical world. If we cannot know this, then we cannot answer the sceptic. Our task, then, is to see whether a contemporary empiricist may have sufficient reason to believe that we hold reliably caused true beliefs about the physical world.

Anti-Nativism and Phenomenalism

Phenomenalism may plausibly be viewed as an attempt to reconcile sceptical arguments with common-sense beliefs. Suppose we accept that all knowledge must be grounded in the data of immediate experience. Suppose, also, that we agree that there can be no knowledge, on this basis, of anything that is independent of experience. Then the phenomenalist should be seen as attempting to accommodate this conclusion within common-sense, by analysing the contents of ordinary beliefs into mere descriptions of patterns within immediate experience itself. This project has proved remarkably difficult to carry through to completion. Indeed, so far as I am aware, no remotely plausible phenomenalist analysis of any proposition about physical objects has ever been provided. But most empiricists have, in one way or another, been committed to just such a programme of analysis.

The reasoning behind the idea that all knowledge must be grounded in immediate experience, which also seems to entail the possibility of a phenomenalist analysis of our concepts, comes partly from the anti-nativism inherent in traditional empiricism. As we saw briefly in chapter 1, if none of our knowledge is innate, and none a priori, then all knowledge must be learned from experience. And if no concepts are innate, and there are no innate domain-specific mental structures embodying information about the world, then the experience in question must be unconceptualized and unstructured. So, from this anti-nativist perspective, the basis of all knowledge would have to be a set of experiences that are not structured in such a way as to contain representations of anything outside themselves—that is to say, a set of sense-data.

Anti-nativists will of course allow that the mind is innately divided into different faculties. They might also allow that the mind has an innate ability to distinguish the differing attributes of our various sense-data, for example recognizing the difference between a sensation of red and a sensation of green, and distinguishing a pain from a tickle.[1] For these abilities do not embody information about anything beyond our sense-data themselves. But for an anti-nativist this would be the full extent of the foundation on which the superstructure of our knowledge would have to rest.

As soon as we accept that our perceptual faculties may be innately structured, however, the picture changes dramatically. For in that case the basis of empirical knowledge would not be sense-data, but perceptions of objects in space outside us. I argued in chapter 6 that our visual faculty, in particular, is innately structured so that incoming information is automatically presented to us in the form of representations of an array of

---

[1] Even Quine, who is in other respects not an ordinary empiricist, seems prepared to allow little more than this. See his concession of innate quality spaces in 'Natural Kinds', in *Ontological Relativity*.

physical objects in three-dimensional space. It will be the beliefs arising out of these representations that form the foundations of our empirical knowledge. It will also be these representations that provide the materials for many of our concepts, perhaps in the manner outlined in chapter 7. Notice that these concepts will constitute modes of classifying physical objects, rather than being abstractions out of, and modes of classifying, sense-data, as phenomenalists would maintain. Indeed, viewed from this nativist perspective, sense-data themselves become entirely marginalized, playing no significant role within the structure of our belief-systems, except in so far as those beliefs happen to be concerned explicitly with our subjective sensations.

It is very likely true that our experience is innately determined as representing a world of physical objects in space around us. Once this is accepted by contemporary empiricists, even as a mere possibility, they immediately lose any motive for attempting a phenomenalist reduction of concepts. For there is then no reason to think that all concepts must be constructed out of sense-data. Indeed, if this form of nativism is correct, then we have found an explanation of why such reductions have proved so difficult to complete. But would this mean that we should also have laid to rest the threat of scepticism? Unfortunately not. For it is one thing to say that our perceptions innately represent to us a world of physical objects outside us. It is quite another thing to say that they *truly* represent such a world. It may be innately determined that some of our beliefs, concerning perceived physical objects, constitute the foundation of our cognitive system, while those beliefs themselves are false. To say that it is beliefs about physical objects, rather than beliefs about sense-data, that are foundational, is not to say that the whole system of our beliefs might not be built upon sand.

To see that this is so, notice that we can have apparent perceptions of objects in space outside us that are illusory. I can seem to see a man lurking in the bushes, while all that is really there is a play of shadows. Indeed, in cases of hallucination, and perhaps also in dreams, we can be presented with apparent perceptions of a three-dimensional reality that is wholly unreal. Moreover, one only needs to recall Descartes's hypothesis of the omnipotent demon to realize that it is at least conceptually possible that all our experience might be just as it is, and all our foundational beliefs about physical reality just as they are, while there is in fact no physical world for those experiences or beliefs to concern. While nativism might help us to avoid a commitment to phenomenalism, it does not, on the face of it, help us to avoid scepticism.

Varieties of Foundation

It might in any case be questioned whether the sort of empiricist nativism sketched above is really coherent, or at any rate consistent with foundationalism. For how can there be foundations to our beliefs that are not completely certain? It might be claimed that to say that a belief is foundational is to say that it does not require support from any other belief. And does this not entail that the belief in question must be certain? For if a foundational belief P were *un*certain, this would surely mean that there must be some other possible belief Q that would lead us to overturn P. But then, it may be claimed, P *does* require support after all, namely from the belief that not Q; which contradicts the hypothesis that P is foundational.

This argument involves a confusion of levels, however. For what is foundational to a cognitive system (a system of first-order beliefs) is not necessarily what is foundational in epistemology, where our concern is to know what we know. We can insist that our perceptual beliefs about the physical world constitute the innately determined foundation of our first-order system of beliefs, since these beliefs are primitive, and are held independently (in general) of our other beliefs. But at the same time we can allow that these beliefs, in being uncertain, do not constitute an appropriate foundation for epistemology. If our concern is to show that our perceptual beliefs about the world constitute knowledge, then we clearly cannot simply take the truth of those beliefs for granted.

It is worth stressing again that the task of epistemology remains essentially the same, whether we endorse justification-alism or reliabilism as a theory of knowledge—namely, to justify our first-order claims to knowledge. In the case of justification-alism this is so because knowledge itself requires justification. But for a reliabilist it will also be true, given that our aim is to know what we know (to achieve second-order knowledge). For the only generally reliable method of forming second-order beliefs concerning which of our first-order beliefs are both true and reliably caused is to find convincing reasons for believing that we have knowledge. (Another point is that the very most that we could hope to obtain in any case, from attempts to combat scepticism by providing reliabilist definitions of the key concepts, would be to shift scepticism from one place to another. Even if scepticism about knowledge, and about justified belief, could be blocked—by virtue of a reliabilist account of these concepts—there would remain scepticism about *internally justified* belief. So there would be no real gain. It is just as challenging to be told that we have available to us no convincing reasons for belief in the physical world, as to be told that we do not have knowledge of such a world.) While many contemporary writers have thought that a shift to a reliabilist conception of knowledge has great significance for epistemology, I disagree, except as regards the possibility of innate knowledge. In particular, we still face the problem of scepticism.

To concede that this is so need not, however, imply a retreat towards phenomenalism. For what we may surely be certain of is that we at least seem to be presented with an array of objects in three-dimensional space. (Of course, to be certain of this, we have to be granted an adequate grasp of the concepts that we use to describe that appearance. But then a similar concession must in any case be granted to phenomenalism. It appears that any enquiry must take for granted that we possess the concepts used in the expression of that enquiry.) There is then nothing to force us to concede that the basis of epistemology must be descriptions of sense-data. Rather, we may take as our starting point that we at least seem to perceive a physical world, and that many of our beliefs at least purport to concern the distribution of objects within that world. Our task is to discover whether we can provide sufficient reason for thinking that most of those beliefs are true, perception itself being a reliable means of acquiring beliefs.

What form may such reasons take? Is our task to *deduce* the existence of the physical world from the existence and nature of our perceptual beliefs? If so, then the task is a hopeless one. For we have already conceded that it is conceptually possible that our beliefs and experiences should be just as they are, while there is in fact no physical reality corresponding to them. This means that there can be no valid argument from the former (our perceptual experience) to the latter (the world). For a valid argument is precisely one for which it is *im*possible that the premisses should be true while the conclusion is false. (And even if there could be such an argument, two problems would remain: first, to give a natural explanation of how reason can possess such power, which is the empiricist challenge; and, secondly, to show why the demon, or the fact that I am dreaming, may not be making me go wrong in my very use of reason.) So if reasons for belief in the physical world had to consist of deductive arguments, then we should be faced with a sceptical conclusion. Taking for granted only that there appears to us to be a physical world, we could not know ourselves to know that there is such a world.

However, why should it be maintained that reasons for belief have to take the form of deductively valid arguments? Why should reasons not include non-deductive arguments, such as inference to the best explanation of a given phenomenon? It is easy to understand why a traditional empiricist should place restrictions on allowable reasons for belief. For such an empiricist is committed to the thesis that all knowledge must be constructed on a twofold basis only: the data of immediate experience, together with analytic truths, including those embodied in deductive arguments. (These latter they understand to be concerned merely with relations between the ideas that we derive from experience.) Hence deductive argument, and only deductive argument, may be taken for granted at the outset as a method for providing reasons for belief.

With nativism allowed to be a possibility, the situation takes on a different aspect. For it may then be that non-deductive principles of belief formation, particularly inference to the best explanation, are an innately given part of our cognitive apparatus. Indeed, we saw in chapter 7 that they very likely are. In which case these principles will count as foundational, alongside our perceptions and perceptual beliefs. But this is not sufficient to show, as yet, that we may make free use of inference to the best explanation in epistemology. One reason for this is that the question whether or not such a principle of inference is innate is an empirical one, needing to be settled by the sorts of consideration adduced in chapter 7. So we cannot know that the principle *is* foundational until we have provided some evidence of it. But such evidence will very likely presuppose that we can have knowledge of the physical world, which is precisely what is in question.

Moreover, the same point can be made here that we insisted on in connection with our perceptual beliefs. To say that a principle of inference is innately given is to say that it belongs to the foundations of our first-order system of beliefs. And if that principle is in fact generally reliable, and the beliefs formed through its use are in fact true, then it will contribute to our first-order knowledge. But it does not follow that we may take that principle for granted when we come to enquire what we may know ourselves to know. Since it is by no means self-evident that inference to the best explanation is a reliable method of acquiring beliefs, there is a question mark over its status when we come to do epistemology, just as there is a question mark over the status of our first-order perceptual beliefs. We have no option but to attempt to justify our reliance on such a principle, even if we know it to be innate.

In the final chapter of this book I shall attempt such a justification, making crucial use of the concession of nativism. But first let us consider whether the principle of inferring to the best explanation could in any case help us to overcome scepticism about the physical world. For if it would not be of any help, then there is little point in discussing whether or not that principle can be justified. Let us therefore proceed for the remainder of this chapter on the assumption that inference to the best explanation *is* justified. This assumption will be vindicated in the chapter that follows.

Inferring to the Best Explanation

What we need to decide, in effect, is whether the hypothesis of our reliable perception of a physical world provides the best available explanation of our perceptual experience, and of the existence of our first-order perceptual beliefs. There are only four candidate hypotheses that we need to consider.

The first is that there is indeed a physical world, and that it is a reliable cause of our perceptual beliefs. According to this hypothesis, the explanation for the fact that there seems to be a room full of people in front of me, and of my consequent belief that there is, is that there really is a room full of people that causes my experience. Let us call this 'the realist hypothesis'.

The second hypothesis also maintains that there is a physical world, but does not suppose that this world is a reliable cause of our beliefs. According to this hypothesis, there may indeed be such things as rooms and people, but the explanation for the fact that I seem to see a room full of people may be some quite unrelated physical situation, such as that some scientists are stimulating the visual centre of my brain, which is in fact floating in a vat of chemicals in their laboratory. Let us call this 'the brain in the vat hypothesis'.

The third hypothesis maintains that there is no physical world, and no cause of our experiences outside our own psychology. According to this hypothesis, our experience is one gigantic hallucination. The explanation for the fact that I seem to see a room full of people will then be some unknown feature of my psychology, which results in me having this experience now, rather than any other one. This might be called 'the dream hypothesis'.

The final hypothesis is again that there is no physical world, but that our experience is caused by some non-physical fact independent of our minds. For example, the explanation for the fact that I seem to see a room full of people may be that there is an all-powerful demon who decides to cause me to have just such an experience. This might as well be called 'the demon hypothesis'.

Which of these available hypotheses provides the best explanation of my experience, and of my perceptual beliefs? Plainly, I think, the realist hypothesis. This can supply a detailed and highly sophisticated explanation of the course of my experience, at the same time warranting subjunctives about what I would experience were I to be differently situated, and counterfactuals about what I would have experienced had I been differently situated in the past. This hypothesis can explain not only the bare fact that I seem to see a room full of people, but also the details of what I seem to see in terms of the details of my physical orientation and the distribution of people in the room. It can also predict that I would seem to see a largely empty corridor if I were to leave the room, and that I would have been seeming to see such a corridor if I had stopped just before I came into the room. So the hypothesis has all the hallmarks of a good explanatory theory: it provides a simple, but powerfully predictive, explanation of the phenomena.

All the other three hypotheses, in contrast, are lamentably weak. The dream hypothesis may be dismissed straight away, precisely because it makes no attempt to explain the details of my experience. The brain in the vat hypothesis, likewise, is incapable of providing detailed explanations. For if my experience of a room full of people is dependent upon the whim of an experimenter, then I cannot explain in detail why I have the experiences that I do. Nor can I predict what I shall experience next, nor say what I might have experienced in other circumstances. The demon hypothesis, too, while being ontologically simple (it only requires us to postulate a single thing outside our minds), is, as it stands, lacking in predictive power. For who knows what the demon might cause me to experience next?

The demon and brain in the vat hypotheses also explain less because they leave the demon's decisions (or those of the experimenter) themselves unexplained. Explanation will come to an end with such claims as that the demon chooses that it should appear to me that there is a room full of people. In contrast, the hypothesis of reliable perception can explain not only why I am having such an experience ('There really are people present'), but also what does the explaining ('They have come to listen to me lecture').

The demon and brain in the vat hypotheses also have less coherence, in that they employ separate principles where the hypothesis of reliable perception can make use of a single set of laws. For example, apparent motion can be explained, on the realist hypothesis, either by movement of the object perceived, or by movement of the perceiver, the same set of physical principles being involved in either case. But on the demon hypothesis, we should have to suppose that there are three distinct principles that the demon chooses to employ. On the one hand he sometimes chooses that it should appear to me that things are in motion. On the other hand, in a situation where he *would* have decided *not* to cause an appearance of motion, if I think to myself 'I shall move', then the demon *does* cause such an appearance. There are also rare cases in which, where he would otherwise have caused an appearance of motion, if I think to myself 'I shall move', then he decides *not* to cause such an appearance (for example, cases where I walk parallel to a moving object in an otherwise darkened room, so that the object appears stationary).

The only recourse for the demon hypothesis (as for the brain in the vat hypothesis) is to suppose that for some reason the demon (or the experimenter) chooses to cause me to have experiences exactly as if there were the sort of physical world I intuitively believe there to be. The only way for the demon hypothesis to have the explanatory and predictive power of realism is for it to be cast in the form 'The demon decides to give me experiences for which the best explanation would otherwise (before mention of the demon) be . . . ', followed by every claim of realism. Then I can explain not only my experience of a room full of people, but what does the explaining: it is because the demon wishes me to believe that there are people who want to listen to me lecture. And I can predict that I should have an experience of a largely empty corridor if I were to leave the room: it is because the demon wishes me to believe that there are corridors that remain in existence when unobserved, and that are generally empty during lecture hours. And so on.

But this now renders the demon hypothesis more complex than realism, since it contains all the latter's complexity, together with the additional supposition of the demon and his decisions. Although the demon hypothesis is ontologically simpler, since it only postulates a single thing outside myself, it is structurally more complex, since every statement within realism will have to appear within the demon hypothesis, preceded by a claim about the demon and his decisions. (This must be so if the latter is to share the former's predictive power.) And it is important to see that in assessing explanations it is structural, rather than ontological, simplicity that matters. This is because, in giving an explanation of some phenomenon, it is always facts rather than objects that do the explaining. The mere existence of rooms and people cannot, for example, explain my experience of a room full of people; it is the fact that there are people in a room within my line of vision that explains. So when we are told that better explanations do not multiply entities beyond what is necessary (which sounds like a direct exhortation to ontological simplicity), this is only because theories that postulate more kinds of thing will also, in general, have to postulate more kinds of fact—but not always, as the comparison of realism with the demon hypothesis makes clear.

Another point is that on the present proposal the demon hypothesis becomes entirely *parasitic* upon realism. The only way actually to work the hypothesis is to forget about the demon, and to get on with the job of constructing the best available version of realism. This means that the demon hypothesis, besides being more complex, also lacks another virtue of a good explanatory theory—that of fruitfulness. One of the marks of a good theory, it is generally agreed, is that it should be fruitful, leading to new predictions that could not previously have been made, and suggesting new lines of research. Yet in this respect the demon hypothesis is entirely inert. It predicts nothing which is not also predicted by realism, and not only does it fail to suggest new avenues of research, but it is impossible to do. research except by forgetting the demon and constructing the best theories that one can within the orbit of realism.

It is clear, then, that the hypothesis of reliable perception provides overwhelmingly the best available explanation of the course of our experience, and of the existence of our perceptual beliefs. (I shall return in the next chapter to the sceptical suggestion that the demon may also be interfering with my powers of best explanation, leading me to reject as worse a hypothesis that may actually be better.) It is therefore a matter of some importance to decide whether we can legitimately rely upon inference to the best explanation in attempting to combat scepticism. If we can, then it would appear that the sceptic can indeed be answered. But before turning to that task, I propose to consider one further objection.

Must Knowledge be Foolproof?

The objection is as follows. Even if it is granted that we may somehow know that inference to the best explanation is a generally reliable method of acquiring beliefs, it is certainly not a foolproof method. On no account of the matter will an inference to the best explanation of some phenomenon provide any sort of guarantee of the truth of the conclusion. Even if it is generally reliable, it is also undeniably sometimes fallible. So, even given that the hypothesis of reliable perception provides the best explanation of the course of our experience, it remains possible that it should be false. Indeed, it remains possible that the demon hypothesis, although worse (in terms of simplicity, fruitfulness, and so on), is in fact the correct one. It may then be objected that our belief that we have knowledge of the physical world cannot, on this basis, constitute knowledge. Since it is not completely certain, we cannot really know it.

This objection represents a perennial temptation in philosophy—namely, to insist that you cannot *really* know anything unless you can be completely certain of it. So even if we have reason for believing inference to the best explanation to be generally reliable, and reason for believing that we have knowledge of the physical world, we cannot properly be said to *know* these things, since we have no absolute guarantee of their truth. On this account, nothing can properly constitute knowledge, except where it is inconceivable that it should be false, given the evidence. (Or perhaps: except where it is inconceivable that it should *turn out to be* false, given the evidence—which is not quite the same thing.)[2]

It is worth noting that we never in fact insist upon absolute certainty in our practical lives, no matter how much may be at stake. For example, imagine yourself as a juror in a murder case, in a state that retains the death penalty, and imagine that the evidence against the defendant is, as we say, overwhelming. Nevertheless, as a scrupulous person, you ask yourself whether you really know that the defendant is guilty. Now suppose that one of the other jurors has proposed an alternative hypothesis—namely, that a group of super-intelligent Martians may have landed on Earth undetected, and planted all the evidence against the defendant for purposes known only to themselves. This hypothesis is certainly a possible one, in the same way that the hypothesis of the omnipotent demon is. It therefore shows that your belief that the defendant is guilty is not absolutely

---

[2] This sort of possibility is explored by Ludwig Wittgenstein in *On Certainty* (Oxford: Blackwell, 1969).

certain. But plainly this alternative proposal would not even cause you to hesitate in handing down a guilty verdict, despite the fact that someone's life hangs in the balance.

What emerges is that it is far from obvious why we should even possess the concept of knowledge, if knowledge really did require complete certainty. For this is not a concept that we have any use for in our practical lives. I suppose it might be replied that such a concept would have a use, as a kind of ideal, to which our ordinary standards of justification may approximate but never attain. But in fact I think that the tendency that philosophers have, to insist that knowledge requires certainty, is based on an illusion, produced by a feature of the concept of knowlege that it shares with many of our other concepts— namely, that it is *purpose-relative*. Let me explain.

Notice that the standards that we ordinarily insist on for a belief to count as knowledge may vary according to the seriousness with which we take the situation in which the question of knowledge arises. Suppose, for example, that I ask you, as a casual enquiry, whether you know that Mary will be at the Philosophy Department party. You reply that you do know this, because she told you that she would be there. In the context, I think we should ordinarily be inclined to allow that you do know that Mary will be at the party. But now imagine an exactly similar exchange taking place in the context of a murder trial. I ask you whether you know that Mary is the murderer, and you reply that you do know this, because she told you so. Even granting your sincerity, it is plain that what you say would be insufficient to convict her. Here, since a life or a life sentence hangs in the balance, we would insist on greater justification than this before allowing that you have knowledge. This sort of variability is easily explained if 'knows' means 'is a true belief that is reliably enough grounded for the purposes in hand'. Moreover, since the contexts in which we employ a concept of knowledge may contain a variety of different purposes and interests of varying importance, it is easy to see that our need for such a concept is fulfilled by the possession of just such a purpose-relative concept.[3]

It is arguable that a great many of our concepts are purpose-relative in precisely this sort of way. Consider, for example, the

[3] For a discussion of the relationship between concepts and human purposes, see my 'Conceptual Pragmatism', *Synthese*, 1987.

term 'flat'. Suppose that you are a farmer, who owns a number of fields that are at present under grass. You are first approached by someone from your local Territorial Army, who is looking for a flat field on which to practise manœuvres. You point to a particular field and say 'That one is flat'. On another occasion you are approached by your local bowls club, who are looking for a flat field on which to play while their bowling green is under repair. You point to the very same field and say 'That one is not flat'. On both occasions, surely, you might say something true. The explanation for this is that 'flat' means 'is flatter than enough things of the kind in question for the purposes in hand'. Where the purposes differ, so does the correct application of the concept.

Now the important point to notice is that, where there are *no* purposes in hand, the standards for applying such concepts are apt to become unlimitedly high. This may be because one then naturally sets standards that would be apt for any *conceivable* purpose. For example, suppose that I ask, in the context of a theoretical discussion of scepticism, where no particular practical project is in hand, 'Is this table really flat?' There would be a very natural tendency to reply in the negative. You would immediately think, for example, that the surface of the table is bound to contain slight irregularities, even if they are undetectable to sight or touch. Thus, with no particular purpose in hand, you are inclined to insist that nothing can *really* be flat unless it is *absolutely* flat.

Similarly, then, in connection with the concept of knowledge. When we ask, outside the context of any practical purpose or project, 'Do we really know that such-and-such?', there is a very natural tendency to deny it. This explains the impulse to reject our claim to have knowledge of the physical world, even given that the hypothesis of reliable perception provides overwhelmingly the best explanation of the course of our experience. But, for all that, our belief that we have knowledge of physical objects may be well enough justified to count as knowledge for any purpose that we might *actually* have. I therefore conclude that, if we can somehow justify our reliance upon inference to the best explanation, scepticism about the physical world will have been decisively refuted.

12

Knowledge by Best Explanation

In this final chapter I shall consider whether the concession of nativism would enable an empiricist to solve the problem of induction, and of justifying non-deductive modes of argument generally.

Preliminaries to a Problem

As I noted in the last chapter, if our perceptual beliefs are in fact both true and reliably caused, then they will constitute first-order knowledge of the world. I also argued that the supposition that they are reliably caused by physical objects is overwhelmingly the best explanation of their existence. Now if we could somehow know that inference to the best explanation is a generally reliable process, then we could also know that our perceptual beliefs are largely true. In that case we should have second-order knowledge also. We should be able to know that we have knowledge of the physical world, thus avoiding scepticism.

A good deal more is at stake in this chapter than this, however. For inference to the best explanation is also implicated in the vast superstructure of belief that we erect on the basis of our perceptual beliefs, including many of the beliefs of common-sense, as well as those of science. For example, many of my beliefs about the past presuppose the reliability of memory. I believe that memory is generally reliable, in turn, because the hypothesis that it is so provides the best explanation of the way in which my memories cohere with those of other people, and with other traces of the past, such as photographs and written records. In the same way, many of my beliefs about the future (such as that my house will be where I expect it to be when I return home in the evening) presuppose that I inhabit a generally regular, stable world, in which most events happen in accordance with predictable patterns. I hold this belief, in its turn, because it provides the best available explanation of the regularities of the past. Even my belief that my chair continues to exist when I turn my back on it is justified, if it is, because it provides the best available explanation for the fact that the chair is still there when I turn around again. And, of course, my beliefs concerning the theoretical entities of science, such as electrons, are held because the hypothesis of their existence provides the best explanation of a variety of experimental data.

All this, too, would be vindicated, if non-deductive modes of argument could be justified. In that case we should not only be able to defeat the sceptic about the physical world. We should also be in a position to defeat the kind of sceptic who, while allowing that we know of the existence of the physical world, and of the objects we are currently perceiving, would deny that we have any knowledge of the past, the future, regions of space not currently under our observation, or such things as electrons, which are too small to be perceived. For in all these cases our beliefs rely, in one way or another, on inference to the best explanation.

I am going to suggest that the concession of a certain sort of nativism may enable the empiricist to justify realism about our knowledge of the physical world. However, one obvious problem in attempting to make use of nativism in combating the sceptic is that most of the arguments given in support of nativism seem to presuppose that there is a physical world, and that inference to the best explanation is a reliable method of acquiring beliefs about that world. In the final section of chapter 7, for example, we concluded that our concept of best explanation is very likely innate, since it does not appear to be explicitly taught, and since it is hard to see how it could be acquired from experience. In effect, we argued that the best available explanation of our possession of the concept *best explanation* is that it is innate. We then immediately face the charge of circularity, if we try to make use of the conclusion of this argument in trying to justify our reliance upon inference to the best explanation. However, I propose to set this problem to one side for the moment. I shall focus initially on the question whether the innateness of our concept of best explanation could in any case help with the problem of justification. I shall return to the charge of circularity in a later section.

Most of the discussions since Hume, of attempts to justify non-deductive patterns of reasoning, have focused upon induction—that is, on arguments whose premisses take the form 'Within our observation all As have been Bs' and whose conclusions have either the form 'Therefore all As are Bs' or the form 'Therefore the next A will be a B'. But, as we noted in chapter 7, induction itself should properly be understood as an instance of the wider category of inference to the best explanation. Accordingly, in the next section I shall briefly outline the traditional problem of induction, and explain the main attempts that have been made to solve that problem. In each case I shall show how those attempts have analogues in the attempt to justify inference to the best explanation. None of these attempts, as they stand, will prove successful.

The Problem of Induction

The first point to notice about inductive arguments is that they are not deductively valid. From the fact that all observed As have, so far, been B, it certainly does not *follow* that all As are Bs. It is always at least conceivable that the very next A we observe, or indeed all the remaining As, will turn out not to be B. Nor does it follow from our observations that *probably* all As are Bs, except in those rare cases where the set of As is finite and we have already observed a substantial percentage of them. So we certainly cannot validly deduce that all As are Bs (or even that the next A will probably be B) on the basis of our past observations. For a valid argument is precisely one where it is inconceivable that the premisses should be true while the conclusion is false.

Suppose we could know that nature is broadly uniform—that the same general laws and principles will obtain in all regions of space and time, and that, as Hume puts it, the future will broadly resemble the past. Then we would know that most regularities in nature are projectable. In which case the insertion of this thesis into inductive arguments, as an extra premiss, would turn them into deductively valid ones. That is, from 'All observed As have been Bs', together with 'Most regularities in nature are projectable', we could validly deduce 'Probably this regularity is projectable, so probably all As are Bs'. The trouble, however, concerns how we are to know that nature is uniform. The only plausible source of our knowledge of this is itself inductive. The only real ground for claiming that nature is mostly regular is that it has mostly been so within our experience in the past. But then the move from 'Nature has mostly been regular in the past' to 'Nature is mostly regular' is an inductive one, in which case our argument will apparently have been moving in circles.

Inductive arguments are not deductively valid, and the attempt to turn them into deductive arguments through the insertion of an extra premiss appears hopeless. Yet they do seem to stand in need of justification. It certainly is not intuitively obvious that generalizing from observed regularities is a reliable way of gaining general knowledge of the world. (And even if it *were* intuitively obvious, empiricists would still demand a natural explanation of the reliability of such intuitions.) If our task is to know what we know, then plainly we cannot simply take the soundness of induction for granted. On the contrary, if we are to make use of induction in epistemology, we shall have to give reasons for thinking that induction is generally reliable. There have in fact been only three main attempts to justify our reliance upon induction. These are the so-called 'pragmatic', 'analytic', and 'inductive' proposals respectively.[1] Only the last two of these need concern us here, for reasons that I shall now briefly explain.

The pragmatic proposal for solving the problem of induction consists in some version of the following argument:

Nature is either regular or not. If it is regular, then induction will gain general knowledge for us better than any other method. If nature is not regular, then *no* method will be successful in acquiring general knowledge. So, either way,

---

[1] For examples of each of these, see the papers collected in R. Swinburne (ed.), *The Justification of Induction* (Oxford: Oxford University Press, 1974).

given that we want general knowledge, it is rational for us to employ induction.

Notice, however, that such an argument does not permit us to conclude that induction is generally reliable. The claim is not that induction is likely to produce truths, but merely that it is more likely to produce truths than any alternative method we might follow.

The pragmatic proposal only establishes, at best, that it is reasonable to *employ* induction, since induction may succeed whereas no other policy will. What we need, however, is an argument whose conclusion is that induction (or rather the wider category of inference to the best explanation) may reasonably be regarded as *reliable*. For what we require is knowledge of its reliability, if we are to use inference to the best explanation to give us knowledge of the physical world. Compare the following: I flip a coin, asking you to call 'Heads' or 'Tails'. I tell you that if you call 'Heads' and guess right I shall give you a prize, whereas if you call 'Tails' and guess right you will get nothing. Plainly you would have good pragmatic reason to call 'Heads', since this is the only strategy that might gain you a reward. But you certainly would not *know* that the coin will come up Heads, even if it does in fact do so. This is exactly analogous to the proposal that we are justified, on pragmatic grounds, in employing induction.

The analytic proposal to solve the problem of induction claims that in order to justify induction we need look no further than our concept of justification itself. On this view, induction forms part of our conception of what justification *is*, in such a way that 'Induction is justified' is an analytic truth. It is therefore a mistake to try to justify induction in terms of anything else, least of all in terms of *de*duction. Rather, inductive arguments form a primitive (that is, basic) part of our practice of justifying and seeking justification for beliefs, fully on a par with deductive arguments. Plainly this proposal admits of extension to the wider category of inference to the best explanation, if anything becoming somewhat more plausible in the process.[2] We need only claim that it is analytic—a necessary component of our concept of justification—that an inference to

---

[2] This view is defended by Armstrong, *What is a Law of Nature?* ch. 4, sect. 5.

the best available explanation of some phenomenon is always justified.

The main trouble with this response to the problem of induction, however, is that it is surely possible for us to question the appropriateness of accepted standards of justification. We can allow that inductive practices form part of our concept of justification, and hence allow that inductively grounded beliefs are justified by accepted standards, and yet ask whether those standards are reliable guides to the truth. The deep question is whether inductively grounded beliefs are likely to be *true*, not whether we *call* such beliefs 'justified'. In order to see this, imagine a community that employs a counter-induction rule. This entitles them to derive, from a premiss of the form 'All observed As have been Bs', the conclusion 'All the remaining As are *not* B'. They, too, may respond to the demand that they justify their practice by claiming that the proposition 'Counter-induction is justified' is, for them, analytic. They may claim, with just as much warrant as ourselves, that the practice of counter-induction forms a necessary ingredient in their concept of justification. Yet plainly their practice cannot be justified (in the sense of 'truth-conducive') if ours is. Exactly similar points may be made in connection with the proposed analytic justification of our practice of inferring to the best explanation.

The inductive proposal to solve the problem of induction argues that, since induction has been remarkably successful in obtaining truths for us in the past, it is, very likely, a generally reliable mode of inference. This argument is, of course, itself inductive, thus immediately giving rise to a charge of circularity. But the proponents of this proposal may reply that the reliability of induction does not actually figure as a premiss of the argument. Rather, induction is here used as a rule, or principle of inference, in the course of arguing for the reliability of such inference. Moreover, we cannot insist that all principles of inference should be transformed into explicit premisses, on pain of a vicious regress.[3] Then, since the only premiss of the above argument is the past success of induction, there is no formal

---

[3] For example, if you refuse to accept the inference from 'P & Q' to 'P' until you are given as an explicit premiss 'if P & Q then P', then you ought equally to refuse to accept *this* inference until you are given as a premiss 'if (if P & Q then P) and P & Q, then P', and so on.

circularity. In the same way, we may defend the wider category of inference to the best explanation by arguing that the best explanation of our past success in the use of such inference is that it is generally reliable as a mode of argument. Here, too, inference to the best explanation is *used*, rather than taken as a premiss.

The situation may be compared with the use of soundness proofs in deductive logic, where we demonstrate (deductively) that our system of rules can only generate truths from truths. Generally such proofs will employ the very same principles of inference that are in question. For example, consider the rule of &-elimination, which allows us to derive 'A' (or alternatively 'B') from 'A & B'. A proof of the soundness of this rule might proceed as follows.

1. If 'A & B' is true, then 'A' is true and 'B' is true (definition).
2. Suppose 'A & B' is true.
3. Then 'A' is true and 'B' is true (from 1 and 2).
4. Then 'A' is true (from 3).
5. So if 'A & B' is true, then 'A' must be true (from 2–4).

Here the &-elimination rule has itself been *used*, in the step from line 3 to line 4. Such proofs are generally said to be *explicative* rather than *persuasive*, in that they cannot convince someone of the soundness of &-elimination who refuses to employ it at all. Rather, they amount to a use of our principles of reasoning to explain to ourselves how it is that such principles can guarantee truth from truth. In the same way, it may be said, we can explicate the reliability of induction, or of inference to the best explanation, by means of an argument that employs that very principle.

The main problem for this proposal arises out of our imagined counter-inductive community, mentioned in connection with the analytic proposal above. For notice that we cannot hope to convince the members of this community of the error of their ways, by pointing to their extensive failures in the use of counter-induction in the past. For they will reason counter-inductively, from the premiss that counter-induction has mostly failed in the past, to the conclusion that most future cases of counter-induction will succeed. Similarly, if we imagine a community of people who have the practice of reasoning to the *worst* explanation of a given phenomenon (either explicitly, or by inverting all our canons of best explanation—for example, preferring the more complex of two explanations, other things being equal), then they will respond to past failures by concluding that counter-explanation is very probably reliable. For the hypothesis of general reliability provides the worst available explanation of their lack of success in the past. The point is that we do seem to need *more* than a merely explicative argument in support of induction or inference to the best explanation. For otherwise what reason do I have, beyond mere laziness, for not switching over to the counter-inductive or counter-explanatory practice?

What emerges is that attempts to justify either induction or inference to the best explanation have not met with any great degree of success. I shall now consider whether we might fare any better if we were given, as a premiss, the innateness of our concept of best explanation. I shall henceforward address myself to this wider category of non-deductive argument, leaving induction to be taken care of implicitly as a result.

The Arguments Revamped

One distinctive feature of the nativism defended in chapter 7 is that it places our use of inference to the best explanation firmly within the faculty of reason. For we do seem to possess a conscious concept of best explanation, which serves to guide our evaluations of competing explanations in particular cases. (This contrasts with the view of Hume, who thought that induction resulted from innately given aspects of our faculty of imagination.) What this then means is that inference to the best explanation does not merely happen to belong within our conception of justification, as the analytic proposal maintains. Rather, it is an innately determined constitutive part of the human reasoning faculty itself. It therefore follows that our imagined counter-explanatory community is impossible for us. For while innateness does not immediately imply inevitability— for example, sexual desire is innate, but some may successfully train themselves not to feel it—inferring to the best explanation is surely too basic a principle of our cognition to be alterable by conscious choice or training. Indeed, many of Hume's reasons for the psychological inevitability of inductive reasoning carry over to the case of inference to the best explanation, now construed as part of our faculty of reason.

We still face the main objection raised against the analytic proposal, however, namely that it cannot settle the question of the connection between accepted standards of justification and *truth*. For this, we need to consider the wider analogue of the inductive proposal (concerning the justification of inference to the best explanation). Here, too, the claimed innateness of our concept of best explanation may be of real help. For if the principle of inferring to the best explanation forms a constitutive part of our cognitive apparatus, then of course we cannot but rely upon it in forming our beliefs, whether at the first-order level, or when we come to do epistemology. If inferring to the best explanation is part of what reasoning *is*, for us, then the fact that it can only be supported by itself need give no particular cause for concern, or for scepticism. Just as our best use of deductive reason supports its own soundness, so too our best use of non-deductive reason supports its own general reliability. For the best explanation of our success in reasoning to the best explanation in the past *is* that such inference is generally reliable.[4] (To see the extent of our past success, reflect again on the way in which inference to the best explanation is implicated in almost every aspect of our practical lives, as well as on its use in science.) Moreover, we cannot now object that counter-explanatory communities could equally well use inference to the worst explanation to support the general reliability of *their* practice. For since such communities are not genuinely possible for us, the mere fact that they are imaginable need not undermine the justification of our own procedure. At the very least, all sense of arbitrariness in the thought that I have not switched over to a counter-explanatory practice is removed.

It appears that an empiricist who can be brought to endorse an appropriate form of nativism, concerning our concept of best explanation, may make considerable headway with the problem of justifying non-deductive modes of inference. In particular, such an empiricist may advance versions of the analytic and inductive proposals that are much stronger than their ordinary —non-nativistic—counterparts. Inference to the best explana-

---

[4] The demonstration that this is so I shall leave as an exercise for the reader.

tion can in consequence be seen to form a constitutive part of human reason, and then the best use of that reason leads to the conclusion that it is itself generally reliable. However, the innateness hypothesis also gives rise to an argument for the reliability of inference to the best explanation that is largely independent of these, as I shall now explain.[5]

A New Argument from Nativism

My claim is that the only plausible evolutionary explanation, of why inference to the best explanation should be an innately determined aspect of human reason, is that it is generally reliable. To this it might possibly be objected that an innate concept of best explanation, while having no survival value in itself, could be a by-product of something that does have value for survival. But this suggestion is too nebulous to take seriously. Until some specific proposal is put forward, concerning what this other innate property might be, it will be more reasonable to believe that our concept of best explanation has survival value in its own right.

A real attack can be mounted, however, on the supposed reliability of at least one of the strands in our concept of best explanation, namely that of simplicity. For it appears easy to explain how a faculty of reason that chooses between theories on grounds of simplicity (other things being equal) could have a value in survival that is unrelated to truth. For theories that are simpler will be easier to operate with and think in terms of, hence making cognitive processing more efficient. I have a number of points to make in response to this suggestion. The first is that, even if one strand in the concept of best explanation bears no relation to truth, this does not begin to undermine the reliability of inference to the best explanation as a whole. Indeed (and this brings me to my second point), it is hard to see what the value of efficient cognitive processing could be, unless it is that we are more likely to be successful in deriving *truths* from simple theories than from complex ones. But then this value will

[5] A precursor of this argument may be found in Konrad Lorenz, *Behind the Mirror* (London: Methuen, 1977). See also the papers collected in G. Radnitzky and W. Bartley (eds.), *Evolutionary Epistemology* (La Salle, Ill.: Open Court, 1987).

only manifest itself, in general, when the theory in question is itself true, or at least close to the truth. While it can indeed happen that true conclusions are derived from wholly false premisses, this will only occur by accident.

But the point I most want to insist on is this. While it is indeed the case that one strand in our notion of simplicity is a matter of notational elegance, or economy of expression, there is another strand in the notion that is arguably more important, and that is best subsumed within the notion of explanatory power. This is the sort of structural simplicity that is at issue when we are told not to postulate entities beyond what is necessary, and to prefer explanations that employ the fewest number of explanatory factors. For example, if my house has been burgled then I should conclude, other things being equal (that is, in the absence of two or more sets of footprints or fingerprints, or of evidence that burglars most often work in pairs), that the crime was the work of a single individual. For the hypothesis of a single burglar is simpler than the hypothesis of two, or three, or four. But the virtue of this sort of simplicity is not a matter of ease of expression. It is, rather, that the simpler theory at the same time explains the *absence* of any evidence of more than one burglar, thus having greater explanatory power than its rivals. And it is hard to see how the survival value of explanatory power can similarly be explained away in terms unrelated to truth.

It can further be objected that while evolution might have been reliable in selecting a concept of best explanation that would generate truths concerning matters of immediate significance for the survival of primitive people, this gives us no reason to think that it will be *generally* reliable, particularly with regard to theoretical science. One response is to deny that there is any principled distinction between common-sense and science. Suppose, for example, that it is said to be distinctive of scientific theories that they deal with unobservable entities, such as electrons. It may then be proposed that evolution gives us no reason to trust inference to the best explanation in connection with unobservables. But we only have to reflect to see that this will not do. For the past is just as unobservable as an electron. Yet being able to reason that the best explanation of fresh paw-marks in the mud is that a tiger has been here very recently may be crucial in ensuring our survival.

Another point is that our concept of best explanation had to be fit to serve us in almost any environment, from the equator to the poles. For of course the distinctive fact about human beings as a species, as we noted in discussing the innateness of folk-psychology in chapter 8, is that we are uniquely adaptable, having a distribution over the globe that no other species has. It is hard to see how any concept could subserve this purpose except one that would be generally reliable. For concepts that are reliable only in connection with phenomena on Earth, for example, can be ruled out on the grounds that they might lead one to predict that the sun will not return next day, and hence to fail to make crucial provision for the morrow.

Of course it must be possible, in principle, to cobble together some sort of artificial concept that would embrace only those aspects of nature with significance for pre-technological survival. And it might then be suggested that evolution could have selected a concept of best explanation that would be reliable in those circumstances only. But to this we can make two separate replies. The first appeals to yet another form of innateness, namely our innate ability to detect genuine resemblances between things. It is plain that we do have such an ability, and that it has survival value in its own right.[6] In which case it is unlikely that pre-technological and scientific phenomena should strike us as so similar (in all those respects that matter for explanation and prediction), when really they are not. The second reply is that we have every reason to think that evolution has delivered us a concept of best explanation whose reliability is *not* limited in the manner suggested. For the best explanation, both of our success in making sense of the world in the past, and of our continuing scientific success, is that our extended use of inference to the best explanation continues to be reliable. (Here the argument merges with the analogue of the inductive proposal, outlined above.)

We can reply in the same sort of way to a rather different objection. This would be that the survival value, and hence reliability, of our concept of best explanation in the past provides no guarantee of continued survival value, or reliability, in the future. For in fact the best explanation for the observed regularity of nature in the past is that nature is governed by laws that operate independently of time and place. Moreover, the

[6] On this see Quine 'Natural Kinds', in *Ontological Relativity*.

best explanation of our success in making sense of the world in the past is that we have an innate ability to detect those properties that figure in projectable laws. So we have reason to think that those properties of the world that conferred survival value on our use of inference to the best explanation in the past will continue to do so. (While there is an obvious circularity in this reply, it is in fact not vicious, but a virtuous coherence, for reasons that will emerge in the next section.)

A final objection to the argument from the innateness of our concept of best explanation to the conclusion of reliability is that such a concept only enables us to make a choice from amongst the various theories that have actually been proposed. So to say that it is reliable is just to say that, of the theories suggested, it tends to select the one that is most likely to be true. But then the use of such a concept will only generally lead to the truth if the correct theory is amongst those antecedently given. So even if the *concept* of best explanation is generally reliable, this gives us no reason to think that *inference* to the best explanation is generally reliable. But the reply to this is easy. It is that we surely have to suppose that our faculty for generating hypotheses would evolve alongside our concept of best explanation. For what would be the survival value of having a concept of best explanation that is reliable in the above sense, if we did not also tend to generate ranges of hypotheses that include the correct one?

I conclude that it is very hard indeed to see what possible advantage possession of a concept of best explanation could confer in survival, unless such a concept were generally reliable in obtaining for us truths, around which we could then construct our plans and projects. In which case, not only does the innateness of inference to the best explanation mean that such inference forms a constitutive part of any possible human concept of justification, but it also gives us two reasons for thinking that there is a connection between our standards of justification and truth. (These reasons are, namely, that the best explanation of our past success in the use of best explanation is general reliability, and that general reliability provides the best explanation of the innateness of our concept of best explanation.) There remains, however, the charge of circularity outlined at the outset of our discussion. For the argument for nativism is itself an inference to the best explanation. Indeed, it is one that presupposes that we have knowledge of a world of physical objects, since we could otherwise hardly be entitled to appeal to evolutionary theory. I shall now turn to the task of rebutting—or at least defusing—this charge.

A Modest Coherentism

Undoubtedly the most important role for the claimed innateness of our concept of best explanation is that it forces on us at least a modest form of coherentism. For recall from chapter 7 that what may make one explanation better than another is, in part, its greater internal consistency and its higher degree of coherence with other received beliefs and theories. In accepting that inference to the best explanation is an innately determined constitutive aspect of human reason, we are therefore accepting that the justification for our beliefs must consist partly in their overall coherence. This may then enable us genuinely to 'bootstrap' our way, not only to a justified belief in the reliability of inference to the best explanation, but also to a justified belief in the physical world as the best available explanation of our experience.

It is true that there is a kind of circle involved in using inference to the best explanation to argue for the innateness of our concept of best explanation, which is then used in turn in arguing that inference to the best explanation is generally reliable. But this circularity is not a vicious one. Rather, it is just what one would expect, given that the relation of justification consists, in part, in coherence within and between neighbouring theories. The best use of our reason leads us to believe that inference to the best explanation is a constitutive part of our reason, and hence that an aspect of what a justification for a belief must consist in, for us, is a matter of coherence. The best use of our reason then leads us to believe that our reason itself is generally reliable, and that the best explanation of our perceptual beliefs is that perception is a generally reliable guide to the states of the physical world. These beliefs and explanations interlock in a highly satisfactory way, enabling us to say that they are amply justified—indeed, that they constitute genuine knowledge.

While we have accepted a coherentist conception of justification, this is still a relatively weak form of coherentism. (Alternatively, it may be considered a weakened form of foundationalism.) For we do not have to allow that *all* coherent networks of true belief constitute knowledge, even provided that there exists no other equally coherent network concerning the same subject-matter. Rather, we can insist that explanatory theories need to be tied down in perception in order to count as justified. So the example given in chapter 5, of a coherent network of beliefs about a character who is in fact fictional, will not count as justified on this view. What is given to us—what must form, as it were, the foundation of our coherent system of belief, if it is to be justified—is that there *seems* to us to be a world of physical objects in three-dimensional space. These seemings of the world constitute the data for which our explanatory theories are constrained to account.

It is not just in accepting a coherentist conception of justification that empiricists should now move away from classic foundationalism. They should also allow an element of coherence to infect those very seemings of physical reality that constitute the foundation for the epistemological enterprise (that is, to know what we know). For when it is pointed out that even such seemings presuppose a grasp of their constitutive concepts (that it seems to you that you are sitting on a chair presupposes that you have a grasp of the concept *chair*), an empiricist need no longer reply (as I did in the last chapter) that grasp of the constitutive concepts must be presupposed in any enquiry. Rather, we can allow that sceptical doubts about our own conceptual abilities are possible, but reply that the best available explanation for the fact that I seem able to make regular judgements and classifications concerning what seems to me to be a largely regular world is that I do indeed possess the ability to classify things in a regular manner.[7] So although my seemings of physical reality constitute a foundation in the theory of knowledge, in so far as they are the basic data that need to be explained, we do not have to claim that they are

[7] See my *Introducing Persons*, ch.6, where this sort of idea is developed in more detail.

absolutely certain, or that they are wholly independent of any other beliefs.

We can reply in a similar manner to the sceptical suggestion that the demon may be making me go wrong in my use of inference to the best explanation, leading me mistakenly to prefer the realist hypothesis over the hypothesis of the demon. For in fact the best explanation of the intelligibility and coherence of my current appraisals of inferences to the best explanation is that such appraisals are generally correct. And again the apparent circularity here is not vicious, but is rather one of virtuous coherence.

Indeed, it may be worth noting that the present line of defence of realism about knowledge may also extend to realism about *truth*, which we discussed briefly in chapter 1, and have since been taking for granted. For the best theory of my abilities to identify and classify things (that is, of the abilities that underpin the representative powers of my thought) is that they are realized in categorical bases in the brain, which operate in accordance with laws of nature that are independent of space and time. So there are determinate truths about how those capacities would respond in remote regions of space and time that are in fact inaccessible to me, which seems to be enough for my present thoughts about those places and times to be determinately true or false. Or so, at any rate, I believe.[8]

I conclude that we are sufficiently justified in believing inference to the best explanation to be a generally reliable method of forming beliefs. This is so because the best explanation of our possession of a concept of best explanation is that the concept is innate, and because the best explanation of its innateness is that it is generally reliable. It is also because the best explanation of our past success in the use of inference to the best explanation is that it is generally reliable, and because the best explanation of past regularities in nature is that nature is law-governed in such a way that past regularities will continue. And it is because the innateness of inference to the best explanation forces us to see that a justification for a belief may consist in the fact that it hangs together with other beliefs in just this sort of interlocking way. We may therefore conclude that

---

[8] For further development of this idea, see my *Metaphysics of the Tractatus*, ch. 15.

we do indeed know that we have knowledge of the physical world, since the hypothesis of reliable perception provides easily the best available explanation of our perceptual beliefs.

An Argument for Open Minds

It is important to be clear about just what the argument above has shown—in particular, that it provides us with no basis for claiming to have defeated someone who is already a convinced sceptic. For even if inference to the best explanation is unavoidable by virtue of its innateness, this does not mean that we necessarily have to believe that the conclusions of such inferences are likely to be true. For example, the analogue of Hume's position on induction remains possible. Someone may concede that inference to the best explanation is unavoidable for practical purposes, while denying that such inference can be justified. And while they may be prepared to act as if the conclusions of such arguments were true—again for practical purposes—they may refuse to allow that they are likely to *be* true.[9] Such a person will happily follow the whole argument of this and the preceding chapter, accepting the conclusions for practical purposes only. So their sceptical position is just as intact at the end as it was at the beginning.

While my argument cannot rationally convince the convinced sceptic, what I do claim is that it should convince all who approach the issue with an open mind. All that is required is that my reader should be prepared to use an inference to the best explanation and believe, at least tentatively, that the conclusion is true. This need not beg any questions in favour of the reliability of inference to the best explanation, since at this stage it can be left as an open question whether or not inference to the best explanation is really reliable. The reader, like myself, may approach the matter perfectly prepared to be convinced that inference to the best explanation is *not* reliable, or to conclude that no verdict can be delivered either way. But in fact, as it turns out, the best use of our faculty of inference to the best

---

[9] For a contemporary writer who adopts just such a position, see Bas van Fraassen, *The Scientific Image* (Oxford: Oxford University Press, 1980).

explanation leads us to the conclusion that such inference is generally reliable.

That the sceptic cannot be defeated does not mean that the sceptic wins. It should come as no surprise that there are epistemological standpoints that are immune to rational persuasion. Consider, for example, someone who is a convinced paranoiac. Of course nothing that we say or do can rationally require them to give up their belief that all are conspiring against them. For anything that we say can immediately be incorporated into the paranoid hypothesis, as being just one more twist in the plot. Yet this need not mean—plainly—that *we* are not justified in believing that there is no plot. What we have to recognize is that the task of epistemology is more modest than we might initially have thought. The task cannot be to produce reasons for belief that will be acceptable to anyone, no matter what their starting point. It can only be to produce reasons acceptable to those who have not yet made up their minds. So while I cannot claim to have defeated the sceptic, I can claim to have sufficiently justified an opposing realism about our knowledge of the physical world.

Conclusion

WHAT has emerged over the last two chapters is that it was a historical accident that many of the classical empiricists should have become attached either to scepticism or to phenomenalism. It was their anti-nativism about concepts and about the contents of experience that led them to phenomenalism. And it was, ultimately, their anti-nativism about concepts (particularly the concept of best explanation) that led them to scepticism about the physical world. Yet, as I argued in chapter 9, opposition to nativism was never really an essential part of the empiricist enterprise. What was crucial was rather opposition to belief-acquisition processes whose reliability cannot begin to be accounted for in natural terms. Nativism was only rejected because a theory of natural selection had not begun to be thought of, and because the only available account of how information or concepts might come to be innate involved intervention by God.

It is, then, a consequence of my position that those later empiricists, such as Russell and Ayer, who continued to oppose nativism and to be attracted towards scepticism or phenomenalism, only did so because they had lost touch with their empiricist roots. They somehow became entangled in the sceptical or phenomenalist projects of classical empiricism for their own sakes, not noticing that those projects had only arisen within empiricism as part of a deeper concern—the concern, namely, that claims to knowledge should be constrained by our best naturalistic theories of the powers of the human mind, and of the mind's modes of access to reality. (Quine, on the other hand, may be seen to have lost touch with the project of classical empiricism in a different way. Instead of requiring merely that epistemology should be consistent with science, he altogether denies the distinction between them. In Quine's version of naturalism, all there is to epistemology is science. But this, implausibly, just rules out of order any attempt to question whether science can really provide us with knowledge.) With the advent of a natural theory of evolution, there was no longer any need for opposition to nativism. And once certain kinds of nativism are admitted as plausible, solutions to many forms of scepticism may drop out quite easily.

It is also a general moral of these last two chapters that much contemporary theory of knowledge has erred, particularly by failing to pay due attention to the traditional question of the possible sources of human knowledge. Many have become ensnared into ever more technical disputes concerning the concept of knowledge, which are largely irrelevant to the epistemological task—to know what we know. They have also tried to solve the problem of scepticism directly, setting aside the issue of the powers and faculties of the human mind (including the question of innateness) as of minor significance. If the strategy of argument I have pursued in the final two chapters has been correct, however, then this concentration of effort has been a mistake. Without considering the general powers of the human mind we cannot hope to settle the question of what we may know ourselves to know. And only if certain forms of nativism are accepted, can we justify realism about our knowledge of the physical world.

If there is one further moral to be drawn from these chapters —as indeed from this book as a whole—it is this. Only by returning to the historical roots of our contemporary debates, and by reconsidering the motives that led philosophers to line up one way or the other, have we found our way to what, in my view, are permanent solutions to the problems of scepticism. Those who do not study the history of philosophy may easily come to mislocate their present position on the landscape, and thus be mistaken about the proper avenues of escape.

The overall conclusion of this book is as I indicated at the outset that it would be—that empiricism is as vibrant and defensible today as it ever was. My contemporary empiricism has been shorn of the traditional opposition to nativism, yet without losing touch with its essentially empiricist concerns. Indeed, there is a strong case for saying that an empiricist should now accept the existence of innate information-bearing mental structures (particularly relating to language and vision), as well as the innateness of at least some concepts and some knowledge. Yet empiricism may still retain its traditional opposition to platonism, and more generally to the idea that we can obtain substantive knowledge by reason alone. In such cases the empiricist demand for a natural explanation, of how our intellects can have acquired the power to obtain for us reliable information about substantive facts independent of our thoughts, remains in full force. Nor does an appeal to natural selection provide a rationalist with any adequate response. The result of these changes is that empiricism may now drop all its historical connections with scepticism or phenomenalism, embracing instead a form of robust realism. My contemporary empiricist may claim (partly via their acceptance of nativism) that we can know ourselves to know that there is a physical world, and that we can know that we have knowledge of many facts about that world, both particular and general.

In the end, an empiricist is what everyone should be.

Further Reading

Classics (available in many editions and translations)

Plato, *Republic* (*c*.380 BC), esp. bks. 6 and 7; *Meno* (*c*.380 BC); *Phaedo* (*c*.380 BC).

Descartes, R., *Discourse on the Method* (1637); *Meditations on First Philosophy*; (1641).

Locke, J., *An Essay Concerning Human Understanding* (1690).

Leibniz, G., *New Essays Concerning Human Understanding* (1704).

Berkeley, G., *Principles of Human Knowledge* (1710); *Three Dialogues between Hylas and Philonous* (1713).

Hume, D., *A Treatise of Human Nature* (1739), bk.ɪ; *Enquiry Concerning the Human Understanding* (1748).

Kant, I., *Critique of Pure Reason* (1781); *Prolegomena to any Future Metaphysics* (1783).

Platonism: For and Against

Benacerraf, P., 'What Numbers Could Not Be', *Philosophical Review*, 1965; 'Mathematical Truth', *Journal of Philosophy*, 1973. (Both are reprinted in P. Benacerraf and H. Putnam (eds.), *Philosophy of Mathematics*, 2nd edn., Cambridge: Cambridge University Press, 1983.)

Field, H., *Science without Numbers* (Oxford: Blackwell, 1980); *Realism, Mathematics and Modality* (Oxford: Blackwell, 1989).

Frege, G., *The Foundations of Arithmetic*, trans. J. Austin (Oxford: Blackwell, 1950).

Hale, R., *Abstract Objects* (Oxford: Blackwell, 1987).

Irvine, A., (ed.), *Physicalism in Mathematics* (Dordrecht: Kluwer, 1990).

Katz, J., *Language and Other Abstract Objects* (Oxford: Blackwell, 1981).

Wright, C., *Frege's Conception of Numbers as Objects* (Aberdeen: Aberdeen University Press, 1983). (2nd Edn., Oxford: Blackwell, 1991.)

Nativism: For and Against

Block, N. (ed.), *Essays in Philosophy of Psychology*, ɪɪ (London: Methuen, 1981), pt. 4.

Chomsky, N., *Language and Problems of Knowledge* (Cambridge, Mass.: MIT Press, 1988), chs. 1, 2 and 5.

Fodor, J., *Representations* (Brighton: Harvester, 1981), ch.10; *Psychosemantics* (Cambridge, Mass.: MIT Press, 1987), ch.1 and epilogue.

Searle, J., (ed.), *Philosophy of Language* (Oxford: Oxford University Press, 1971), pt. VII.

Stich, S., (ed.), *Innate Ideas* (Berkeley, Calif.: California University Press, 1975).

The Nature of Knowledge

Bonjour, L., *The Structure of Empirical Knowledge* (Cambridge, Mass.: Harvard University Press, 1985).

Craig, E., *Knowledge and the State of Nature* (Oxford: Oxford University Press, 1990).

Dancy, J., *Introduction to Contemporary Epistemology* (Oxford: Blackwell, 1985), chs.2–4, 8, and 9.

Goldman, A., *Epistemology and Cognition* (Cambridge, Mass.: Harvard University Press, 1986), pt. 1.

Lehrer, K., *Theory of Knowledge* (London: Routledge, 1990).

Nozick, T., *Philosophical Explanations* (Oxford: Oxford University Press, 1981), ch. 3.

Scepticism: For and Against

Bonjour, L., *The Structure of Empirical Knowledge* (Cambridge, Mass.: Harvard University Press, 1985), ch.8.

Dancy, J., *Introduction to Contemporary Epistemology* (Oxford: Blackwell, 1985), chs.1, 5, 9, 10–13.

Hookway, C., *Scepticism* (London: Routledge, 1990).

Lycan, W., *Judgement and Justification* (Cambridge: Cambridge University Press, 1988), pt. 2.

Stroud, B., *The Significance of Philosophical Scepticism* (Oxford: Oxford University Press, 1984).

Swinburne, R., (ed.), *The Justification of Induction* (Oxford: Oxford University Press, 1974).

Unger, P., *Ignorance: A Case for Scepticism* (Oxford: Oxford University Press, 1975).

Index